Microsystems and Nanosystems

Series Editors

Roger T. Howe, Stanford, CA, USA

Antonio J. Ricco, Moffett Field, CA, USA

Building on the foundation of the MEMS Reference Shelf and the Springer series on Microsystems**, the new series Microsystems and Nanosystems comprises an increasingly comprehensive library of research, text, and reference materials in this thriving field. The goal of the Microsystems and Nanosystems series is to provide a framework of basic principles, known methodologies, and new applications, all integrated in a coherent and consistent manner. The growing collection of topics published & planned for the series presently includes: Fundamentals • Process Technology • Materials • Packaging • Reliability • Noise MEMS Devices • MEMS Machines • MEMS Gyroscopes • RF MEMS • Piezoelectric MEMS • Acoustic MEMS • Inertial MEMS • Power MEMS • Photonic MEMS • Magnetic MEMS Lab-on-Chip / BioMEMS • Micro/nano Fluidic Technologies • Bio/chemical Analysis • Point-of-Care Diagnostics • Cell-Based Microsystems • Medical MEMS • Micro & Nano Reactors • Molecular Manipulation Special Topics • Microrobotics • Nanophotonics • Self-Assembled Systems • Silicon Carbide Systems • Integrated Nanostructured Materials Structure • A coordinated series of volumes • Cross-referenced to reduce duplication • Functions as a multi-volume major reference work **To see titles in the MEMS Reference Shelf and the Springer series on Microsystems, please click on the tabs "Microsystems series" and "MEMS Reference Shelf series" on the top right of this page.

More information about this series at http://www.springer.com/series/11483

Kirill Poletkin

Levitation Micro-Systems

Applications to Sensors and Actuators

 Springer

Kirill Poletkin 🆔
Innopolis University
Innopolis, Russia

ISSN 2198-0063   ISSN 2198-0071   (electronic)
Microsystems and Nanosystems
ISBN 978-3-030-58910-3   ISBN 978-3-030-58908-0   (eBook)
https://doi.org/10.1007/978-3-030-58908-0

This Springer imprint is published by the registered company Springer Nature Switzerland AG
The registered company address is: Gewerbestrasse 11, 6330 Cham, Switzerland

Acknowledgements

This work combines the results related to research and development in the field of inductive levitation micro-systems obtained by the author jointly with his co-authors from different universities and countries such as Moscow Aviation Institute (Russia), Nanyang Technological University (Singapore), Freiburg University and Karlsruhe Institute of Technology (Germany) during the author's postdoctoral journey taken over the last 10 years. The author thanks...

- *Jan Korvink* as a co-author and Habilitation supervisor for sustainable maintenance of efforts of the author in the topic of electromagnetic simulation;
- *Ulrike Wallrabe* for hosting author's project funded by the Alexander von Humboldt foundation and supporting continuously his research;
- *Christopher Shearwood* for introducing the field of micromachined inductive levitation systems and inspiring the author to carry out research in this field and to write his very first articles about this topic;
- *Alexandr Chernomorsky* for his continuous support of research conducted by the author and many fruitful discussions about electromagnetic levitation, which were then transformed into interesting research problems presented in a number of articles;
- colleagues and co-authors such as *Zhiqiu Lu, Asa Asadollahbaik, Ronald Kampmann, Ali Moazenzadeh, Dario Mager, Sagar Wadhwa* and others.

The author deeply thanks *Manfred Kohl* and *Ulrike Wallrabe* as the coordinators for maintaining his current project in the framework of the priority German Research Foundation programme SPP 2206/1. The main part of the presented results was obtained under the financial support from Alexander von Humboldt Foundation and Deutsche Forschungsgemeinschaft (DFG).

Contents

Chapter 1
Introduction to Levitation Micro-Systems

Levitation is a physical phenomenon, which is associated with a body floating freely without mechanical attachment with a surrounding. This fascinating approach to suspend a body can be realized by utilizing several known methods including a jet of gas, intense acoustic waves and others. However, only the levitation using electric and magnetic force fields so-called *electromagnetic levitation* can avoid reliance on an intermediate medium to act on a floating body.

Electromagnetic levitation can be defined as the stable or indifferent equilibrium of a body without material contact to its surrounding environment, which is reached by means of stored electromagnetic energy inducing, in turn, the *ponderomotive forces* acting on the levitated body and equilibrating gravitational forces. Consequently, it allows performing true levitation in a vacuum environment [1]. This feature means that the friction can be controlled or eliminated so that a levitated body floats frictionlessly.

Electromechanical systems implement the principle of electromagnetic levitation on a micro-scale is referred to as electromagnetic levitation micro-systems or *levitation micro-systems*. Levitation micro-systems offer a fundamental solution to overcome the domination of friction over inertial forces at the micro-scale [2]. The achievable motion range of a levitated body is considerably extended, thereby yielding micro-systems with wider operational capabilities, and at the same time, significantly reducing the dissipated energy. For some critical applications of micro-systems in harsh environments (with vibration, heat and chemical influences), electromagnetic levitation can prevent contact of a levitated micro-object with harmful surfaces. Besides, the lack of contact eliminates mechanical wear, thus ensuring a long device lifetime.

These circumstances have already inspired many researchers and scientists to develop and demonstrate levitation micro-system devices with higher performances, which have found wide applications in micro-inertial-sensors, micro-motors, switches, micro-accelerators, particle traps, digital micro-fluidics, nano-force sen-

K. Poletkin, *Levitation Micro-Systems*, Microsystems and Nanosystems, https://doi.org/10.1007/978-3-030-58908-0_1

sors, micro-transporters, micro-bearings, etc. Recent achievements in the technology of permanent micro-magnets [3], 3D micro-coil fabrication [4, 5], and 3D micro-coils with integrated magnetic polymers [6], have drastically increased the efficiency of levitation micro-systems, and predict promising future applications in areas such as robotics, physics, optics, chemistry, medicine and biology.

1.1   Levitation Micro-Systems. Classification

A levitation micro-system (L-micro-system) consists of two primary elements, namely, a set of sources of force fields, and a micro-object (or a set of micro-objects) which is levitated by these fields in a manner such that all six degrees of freedom of the micro-object (or each micro-object in the set) is stable. The main issue in employing levitation in a micro-system is related to stability or stable levitation, which thus provides a means for the classification of L-micro-systems. Stable levitation signifies that a levitated mass is restored to rest when it is slightly displaced in any linear and angular directions from an equilibrium position. According to the Earnshaw theorem [7], the stable levitation of a body located in a static force field, in which the generated force acting on the body is dependent on distance by the inverse square law, is impossible. The force fields produced by systems of charged particles and magnetic dipoles are the simplest examples of such systems, in which the condition of stable levitation is no feasible. In other words, the potential energy (stored energy) of such systems does not possess a local minimum at the equilibrium state.

However, linear and angular displacements of the levitating object can be measured by sensors. Thus, received signals from sensors can be used to change the energy flux generated by the force field. Hence, choosing the right law of control (closed-loop control), stable levitation in systems having inverse square force laws becomes possible. Stable levitation based on closed-loop control is called *active levitation*. A static magnetic field can imitate the control system if the levitated mass is, for instance, diamagnetic or superconducting. Also, using dynamic phenomena such as, for instance, a resonant circuit driven by the ac voltage source and oscillatory electric fields, both conducting micro-objects [8] and charged particles [9] can be stably levitated. The levitation mode that overcomes the restrictions of the Earnshaw theorem, without using closed-loop control, is called *passive levitation*.

The use of either active or passive levitation has its advantages and disadvantages. Nevertheless, the principal difference between both manners is that the first one requires displacement sensors to control all motion of the levitated mass, and the latter is not constrained in the same way. Both manners are widely employed in L-micro-systems. Using this feature, and depending on a source of force field, L-micro-systems can be further classified as [10]:

- *Electric levitation micro-systems*: force interaction between dielectric and conducting materials located into the electric field;

- *Magnetic levitation micro-systems*: force interaction between magnets and direct current (dc) or magnetic materials and dc and also including

  - *Diamagnetic levitation micro-systems*: force interaction between diamagnetic materials and magnet;
  - *Superconducting levitation micro-systems*: force interaction between superconducting materials and magnet;
  - *Inductive levitation micro-systems*: force interaction between conducting materials and alternating current (ac);

- *Hybrid levitation micro-systems*: a combination of different force fields.

1.2 Electric Levitation Micro-Systems

In 1990, the electric levitation micro-system (EL-micro-system) based on passive levitation was proposed by Kumar et al. as a solution of the friction problem in micro-actuators [8]. Using a tuned LCR (L = inductance, C = capacitance, R = resistance) circuit, necessary conditions for open-loop stable levitation of a micro-disc and its dynamics were analysed. The authors analysed two designs having single and dual stators (see Fig. 1.1a, b) and considered their potential application in micro-motors and bearings. Then, in 1992, the single-stator design for an application as micro-bearing was successfully experimentally studied by levitating a micro-plate having a thickness of 180 μm [11]. The actuator was able to generate a force up to 0.2 μN along the vertical axis. Then, using a similar single-stator design of EL-micro-system as shown in Fig. 1.1a, but with an active approach for levitation, Higuchi's group from the University of Tokyo electrically suspended an aluminium disc [12] and a glass plate [13]. Later, adding a variable-capacitance actuation principle to the actuator based on this single-stator design, the same group suspended and linearly transported a silicon wafer [14], and demonstrated a motor in which a rotor was revolved with a speed of 60 rpm by applying a rotational torque of 8.6 μNm [15].

In 1994, SatCon Technology Co. developed and successfully demonstrated a prototype of a rotating micro-gyroscope as proof of concept [16]. In this prototype, a

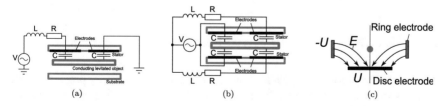

Fig. 1.1 Electric levitation micro-systems based on the passive levitation: **a** single-stator design; **b** dual-stator design; **c** levitation of a charged particle in a hyperbolic oscillating electric field E (U is the applied electric potential)

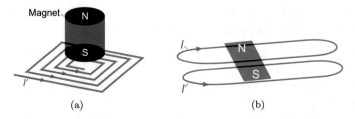

Fig. 1.2 Magnetic levitation micro-systems (I is the electric dc current): **a** square coil design for micro-assembly [32]; **b** set of coils for a linear actuator [33]

rotor disc having a diameter of 200 μm was driven by multi-phase radial torque actuation. Employing an active approach for stable levitation, the gap between the levitated disc and surrounding electrodes was 2 μm. The small distance made it possible to generate a drive torque of 10×10^{-11} N m by applying a nominal drive actuator potential of 5 V. This successful concept found applications in micro-motors [17–19], gyroscopes [20–23], multi-inertial sensors [24, 25] and accelerometers [26–29], and was further improved by other research groups in terms of micromachining fabrication and signal conditioning. As a result, in 2005, an electrostatic micro-motor which was able to rotate a 1.5 mm diameter ring-shaped rotor having a speed up to 74,000 rpm was demonstrated by Nakamura [25]. Using Ball Semiconductor technology, Toda et al. electrostatically levitated a Si-sphere with a diameter of 1 mm and applied it as a three-axis accelerometer [26].

By using an oscillatory electric hyperbolic field (the scheme is shown in Fig. 1.1c), Post et al. passively trapped a charged gold particle of 7.5 μm diameter [30]. Discussing the potential performance of the prototype for application as an inertial sensor, it was shown how trapping parameters could be adjusted to increase dynamic range and demonstrated that a sensitivity comparable to that of a commercial MEMS device was achievable.

An electric micro-actuator based on active levitation found application in high-temperature containerless materials processing. Rhim et al. used this technique to suspend a zirconium sphere of 2.5 mm diameter in vacuum, and demonstrated multiple superheating-undercooling-recoalescence cycles in the sample (heating the sample up to 2270 K) [31].

1.3 Magnetic Levitation Micro-Systems

A magnetic levitation micro-system (ML-micro-system), proposed for micro-assembly, was presented by Ye et al. in work [32]. The fabricated prototype was composed of three layers of silicon and a permanent magnet and fabricated using micromachining technology. The top layer consisted of a groove and many pairs of small square coils which were used to generate driving forces. The middle layer consisted of a groove which comprised the moving guide of a permanent magnet. A long rectangular coil built on the bottom layer, as shown schematically in Fig. 1.2a, was used to generate the levitating force. The permanent magnet of $1 \, mm^3$ volume was lifted to 200 μm by generating a levitation force of 20 μN. Dieppedale et al. developed and demonstrated a prototype of ML-micro-actuator, in which a levitating mobile magnet of $5 \times 40 \times 100 \, \mu m^3$ volume was used [34]. The prototype employed a bistable actuation principle and was used as a micro-switch. A linear actuator based on active levitation was fabricated and experimentally studied by Ruffert et al. [33, 35]. Using a set of parallel coils as schematically shown in Fig. 1.2b, with the length of the magnetic guide coils of 11.4 mm, a permanent magnet having a length of 4 mm was levitated at a height of 100 μm by generating a repulsive force of 1.5 mN.

In 2005, Dauwalter and Ha described the novel concept of a magnetically suspended spinning rotor [36]. The concept was based on a novel, patented, planar magnetic actuator and position sensor configuration, with a rotor that is rotated by a multi-phase electromagnetic spin motor. The authors proposed its potential application as multi-inertial sensors and micro-motors.

1.4   Diamagnetic Levitation Micro-Systems

In 2004, Lyuksyutov et al. reported the application of diamagnetic levitation micro-systems (DL-micro-systems) for on-chip storage and high-precision manipulation of levitated droplets of pico to femtoliter volume [37]. The developed magnetic micro-manipulation chip consisting of a core with two neodymium-iron-boron permanent magnets (as shown in Fig. 1.3a) having a size of 10 mm $\times$ 10 mm $\times$ h, where the height h was in the range of 100–2000 μm, was mounted on a steel substrate containing a set of electrodes. Droplets were moved by current pulses. For measurements, a typical potential well of around 100 μm in width and glycerin/water (15% glycerin by volume) mixture droplets of $(6 \pm 0.3) \, \mu$m, $9 \pm 0.3 \, \mu$m and $(14 \pm 0.3) \, \mu$m in diameter, were used. Lyuksyutov et al. showed that the stiffness of the restoring force for a droplet of $6 \pm 0.3 \, \mu$m was around (3.0 ± 0.6)fN/μm. Using electroplated CoPt micro-magnet arrays, Chetouani et al. achieved the levitation and trapping of 3 μm -large diamagnetic latex beads and microdroplets of water, ethanol and oil [38].

A new technology to handle low-volume droplets without any physical contact was presented by Haguet et al. in 2006. This innovative approach of digital microfluidics is based on the diamagnetic trapping of liquids above permanent NdFeB macro- and micro-magnets [45, 46]. Later, Pigot et al. fabricated an array of micro-magnets of length 1350 μm, having a vertical polarization of 1.2 T and demonstrated the stable levitation of bismuth particles of 5 μm radius in air, and copper and silicon particles in paramagnetic water [39]. In [40], Kauffmann et al. proposed an integrated electromagnetic hybrid device combining dielectrophoretic (DEP) forces to control the position of levitating droplets along a magnetic groove. The use of these electrodes as micro-conductors to produce variable magnetic fields for magnetophoretic droplet actuation was considered and compared to dielectrophoretic actuation.

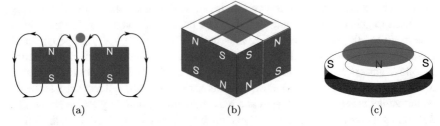

Fig. 1.3 Diamagnetic levitation micro-systems: **a** two magnet design for levitation of droplet/partical [37–40]; **b** four magnet design for levitation of a plate [41, 42]; **c** the design with a ring and disc shaped magnets [43, 44]

Garmire et al. considered the application of DL-micro-systems as inertial sensors and demonstrated a proof-of-concept of a diamagnetically levitated micro-accelerometer, in which a plate consisting of a pyrolytic graphite and silicon layer of 6 and 20 μm thickness, respectively, was levitated by a set of four micro-magnets with the design shown in Fig. 1.3b [41]. Using a similar magnet design, Ando et al. demonstrated an inductive readout to measure the linear displacement of a levitated square graphite plate [42]. In 2015, Su et al. levitated a 600 μm -thick micro-disc, which was micro-patterned from highly oriented pyrolytic graphite as a strongly diamagnetic material, to a height of 132 μm by using a ring and disc-shaped magnet design as shown in Fig. 1.3c. The magnet had an outer diameter of 3.175 μm and an inner diameter of 1.588 μm [43, 44]. The application of DL-micro-systems in nano-force sensors was studied theoretically and experimentally by Abadie et al. in [47], where the authors demonstrated a prototype of their sensor with a bandwidth of 4 Hz and a resolution of 5 nN.

Using design shown in Fig. 1.3a, X. Chen et al. levitated pyrolytic graphite plate above four cubic Nickel-coated NdFeB magnets in a vacuum chamber over a pressure range of 1×10^{-4}–1000 mbar [48]. Also, the authors fabricated an electrode structure between the levitated plate and magnets to excite lateral, angular and vertical modes of oscillation to characterize the proposed DL-micro-system as the levitated resonator. It was experimentally demonstrated a Q-factor of 400 and 150 corresponding to angular and vertical mode of oscillation, respectively, for a plate length of 4 mm at room temperature and low pressure of 0.001 mbar. The authors predicted that if the length of plate is reduced to micro-scale, a Q-factor can be dramatically increased to be larger than 100 millions.

1.5   Superconducting Levitation Micro-Systems

In 1989, Kim et al. utilized magnetic levitation using the Meissner effect and presented a new type of superconducting micro-actuator of the order of 100 μm in size [49, 50]. The slider and the stator of the actuator consisted of linear arrays of vertically magnetized permanent magnet and superconductor strips. The authors showed that when the slider is levitated at 50 μm upon maintaining the operating temperature at 90 K maximum values of driving and levitating forces were 0.26 and 1.7 N m^{-2} A^{-2}. The movement of the slider in a particular direction (right or left) was obtained by choosing the driving mode. Coombs et al. presented a micro-motor, where the rotor was levitated by superconducting bearings [51]. The bearings were self-positioning, relying on the Meissner effect to provide a levitation force which moved the rotor into position, and flux pinning to provide stability thereafter.

1.6 Inductive Levitation Micro-Systems

The first micro-machining prototype of inductive levitation micro-systems (IL-micro-systems) was pioneered and fabricated by Shearwood et al. [52] in 1995. A two-coil design, comprising a nested stabilization and levitation coil as shown in Fig. 1.4a, was fabricated by using surface micromachining technology on a soft magnetic backing plane and was able to stably levitate an aluminium disc having a diameter of 400 μm and thickness of 12 μm. Shearwood et al. demonstrated the actuation by moving the disc along the vertical direction within a range of 30 μm and proposed a potential application of the prototype in micro-inertial sensors, micro-motors and actuators [55]. Then in 1996, modifying the two coil design into a three coil stator, the Shearwood group rotated the disc with a speed of 100 rpm [56]. In the framework of the same work, evaporating a gold layer on the one side of the Al disc and then using such a multi-layer proof mass as a micro-mirror, the Shearwood group showed the application of IL-micro-systems in adaptive optics. Later in 2000, Shearwood et al. improved the original design of the IL-micro-system by embedding a four-phase stator. As a result, in addition to vertical actuation, rotation of a micro-disc with a speed of 1000 rpm was demonstrated [57]. In 2006, Zhang et al. separately fabricated coils for levitation and rotation in one IL-micro-system prototype, and was able to rotate a disc of a diameter of 2.2 mm with a speed of 3035 rpm at a height of 200 μm. The main disadvantage of the above-mentioned prototypes was a very high operating temperature, rising to 600 °C [57] due to the planar design of the micro-coils. Proposing a spiral coil design as shown in Fig. 1.4b, Tsai et al. fabricated a prototype that was able to stably levitate a conducting cap-shaped proof mass with an operating temperature of 100 °C [53].

In 2010, the Wallrabe and Korvink groups from the University of Freiburg jointly developed 3D micro-coil technology [4, 5], which was later utilized to fabricate an IL-micro-system prototype based on a design of two solenoidal micro-coils (see Fig. 1.4a) with a drastic reduction in heat dissipation [58]. In the framework of this prototype, Wallrabe's group demonstrated stable levitation of a disc having a diameter of 3.2 mm and a thickness of 25 μm, as well as a large increase in actuation range along the vertical axis up to 125 μm by changing the ac current in the coils within the range 80–100 mA. This design was further intensively studied and characterized in terms of stability, dynamics and electrical performance [54, 59–62]. In particular,

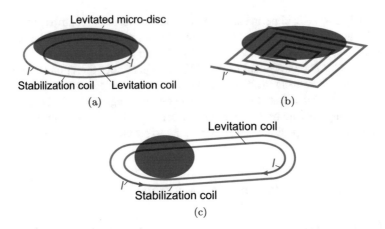

Fig. 1.4 Inductive levitation micro-systems: **a** two coils design, I is the electric ac current [52]; **b** spiral coil design [53]; **c** two racetrack-shaped solenoidal 3D wire-bonded micro-coils [54]

recently Institute of Microstructure Technology presented a new IL-micro-actuator design based on 3D micro-coil technology, where the coil structure consists of two racetrack-shaped solenoidal 3D wire-bonded micro-coils, to be used as Maglev rails as shown in Fig. 1.4c [54]. The design allowed an additional translational degree of freedom to the levitated micro-object and was proposed to be applied as a micro-transporter. A prototype was fabricated and experimentally studied. It was showed that micro-objects with different shapes, such as disc-shaped and plate-shaped platforms could be stably levitated and transported. In 2016, a prototype with integrated polymer magnetic composite core demonstrated the lowest operating temperature of around 60 °C among all previously reported IL-micro-systems [63] and indicated the potential for further decrease towards ambient temperatures [64, 65].

1.7   Hybrid Levitation Micro-Systems

An attractive application of L-micro-systems is that they provide one of the best mechanisms with which to handle very small and fragile objects, for example, when the force required for handling is limited by thermally fluctuating forces due to Nyquist and Johnson noise. Two ways for guiding the level of fluctuating forces, which also define the limit of performance of L-micro-systems, have been recognized and implemented. The first technique decreases the friction, μ, in the L-micro-system by controlling vacuum, since the fluctuating force due to Nyquist noise, $F_N = \sqrt{4k_B T \mu \Delta f}$, where k_B is the Boltzmann constant, T is the absolute temperature and Δf is the bandwidth, is thereby directly controlled. But the Johnson noise force, $F_J = \sqrt{4k_B T c}$, where c is the stiffness, limits the possibility to control fluctuations only utilizing a vacuum. It requires to control the dynamics of the system as well, which can be achieved by a combination of different force fields. This fact brings us to extend our list of classifications of L-micro-systems by the term *hybrid levitation micro-systems* (HL-micro-systems).

HL-micro-systems based on a combination of electric and inductive force fields can control the vertical component of the stiffness of an inductive suspension. A quasi-zero stiffness was proposed by Poletkin et al. [66]. The concept was realized in a micro-prototype reported in [67, 68], where electrostatic forces acting at the bottom and top surfaces of the inductively levitated disc maintained its equilibrium position, and, at the same time, decreased the vertical component of the stiffness of the inductive suspension by increasing the strength of the electrostatic field. In 2015, Poletkin et al. presented an HL-micro-actuator based on the same combination of force fields, where linear and angular positioning, bistable linear and angular actuation and the adjustment of stiffness components, of a levitated disc having a diameter of 2.8 mm and a thickness of 13 μm, were implemented [69].

Moreover, HL-micro-systems based on a combination of electric and inductive force fields found applications as gyroscopes, accelerators, micro-mirrors and transporters. Thus, Tsai et al. inductively levitated a conducting cap-shaped rotor, and electrostatically revolved it with a speed of 2000 rpm [53]. Subsequently, Kraft's group proposed design of the accelerator and experimentally demonstrated that a 1 × 1 mm^2 sized and 7 μm thick micro-object could be levitated to a maximum height of 82 μm and propelled forward continuously at a maximum velocity of 3.6 mm/s [70]. Later in 2019, using the same combination of force fields, Kraft's group proposed a design of a contactless electromagnetic levitation and electrostatic driven micro-system, in which a fabricated aluminium micro-rotor can incorporate a mirror and control it to rotate around the vertical axis within the range of ±180°, which enlarges the scanning angle dramatically compared with conventional torsion micromirrors [71]. In 2016, Poletkin et al. demonstrated a prototype of a micro-transporter employing 3D micro-coil technology and electrostatic actuation to propel a levitated micro-object [62].

Combining diamagnetic and electric force fields, Kauffmann et al. successfully levitated several repulsively charged picoliter droplets, performed in a ∼1 mm^2 adjustable flat magnetic well, which was provided by a centimetre-sized cylindrical permanent magnet structure [73]. The application of the HL-micro-system using the same combination in micro-motors was exhibited by Liu et al. [74] and Xu et al. [75].

In 2009, Liu et al. presented an HL-micro-system, where diamagnetic and inductive force fields were combined [76]. Such a hybrid provides an improvement in the energy performance of the levitation micro-system in compared with that based on only diamagnetic or inductive effect.

1.8 Future Trends

In order to employ the phenomenon of electromagnetic levitation in micro-systems, a wide spectrum of physical principles have been utilized and successfully implemented by using different techniques of micro-fabrication. This fact already gave rise to a new generation of micro-systems, which have found application in a widespread domain as summed up and shown in Table 1.1. However, the potential and the resulting capabilities of L-micro-systems have not yet been exhaustively demonstrated, mainly due to a number of remaining technological challenges.

Electric levitation micro-systems have been well established from the micro-fabrication point of view. However, accurate control of the gap, between a levitated mass and surrounding electrodes, is required to be efficient. Solving this issue by employing the active control for stable levitation increases complexity for designing EL-micro-systems, which in turn restricts the possibility of improving their performance for the current application and a domain of their further application.

Using a permanent magnet in magnetic, diamagnetic and superconducting levitation micro-systems has a manifest benefit, such as passive levitation without the requirement of direct energy consumption. However, magnetic levitation employing static magnetic fields either requires new diamagnetic materials with a high magnetic susceptibility ($>1 \times 10^{-4}$), or novel high-temperature superconducting materials. The development of effective fabrication techniques for the integration of permanent micro-magnets is also required.

The recent improvement in the performance of inductive levitation micro-systems by utilizing 3D micro-coil technology together with the integration of a polymer magnetic composite material for flux concentration has established them as a very promising direction for the further development and the application of inductive levitation in L-micro-systems. Nevertheless, further development of the magnetic properties of electroplated soft magnetic layers is needed.

Another point is that levitation itself will benefit from improvements in the performance and capabilities of L-micro-systems through a combination of different force fields (hybrid levitation). A key potential feature of HL-micro-systems is to use passive levitation. In particular, inductive levitation produces a passive levitation effect, which can be combined with all other force fields without interrupting the stability. This fact leads to new combinations. For instance, inductive levitation of a permanent micro-magnet interacting with the dc electric current will improve the energy performance of L-micro-systems. A combination of an inductively levitated micro-object with an optical force field for revolution and transportation will improve the functionality of the L-micro-system. Other combinations of force fields are of course also possible.

Hybrid levitation micro-systems open a new perspective for developing micro-systems with the increasing demands on functionalities to meet emerging challenges in innovative approaches to control different physical processes occurring, for instance, in mechanics, optics and fluidics at a micro-scale. Such micro-systems lead to implement multidisciplinary and innovative approaches based on multi-stable

Table 1.1 Levitation micro-systems

Type	Sources of force field	Materials of levitated mass/rotor	Applications	Technologies	References
Active					
Electric	Electrical field	Dielectric, semiconductors and conductive material	Suspension	Bulk	[12–14, 31]
			Accelerometer	Exotic Ball-micromachining	[26, 27]
			Motor	Bulk-micromachining	[15, 17–19]
			Gyroscope		[16, 20–23]
			Multi-sensor		[24, 25]
			Acelerometer		[28, 29]
Magnetic	DC current	Magnet	Actuator	Surface-micromachining	[32–35]
		Iron	Motor	Milli-machining	[36]
Passive					
Electric	Electrical field	Conductive material	Suspension	Surface-micromachining	[11]
		Particles	Trapping	Mix of micromaching and Bulk	[30]
Inductive	AC current	Conductive material	Actuator		[52, 53]
			Gyroscope	Surface-micromachining	[57, 72]
			Motor		
			Suspension		[55, 56]
				3D-micro-coil	[58–61]
					[63–65]
Diamagnetic	Magnet	Diamagnetic	Trapping, Digital micro-fluidic	Mix of Surface-micromachining and Bulk	[37–39] [40, 45, 46]
			Accelerometer		[41, 42]
			Gyroscope		[43, 44]
			Force-sensor		[47]
			Resonator		[48]

(continued)

Table 1.1 (continued)

Type	Sources of force field	Materials of levitated mass/rotor	Applications	Technologies	References
Supercon-ducting	Magnet	Superconductor	Micro-bearing	Surface-micromachining	[50, 51]
Hybrid					
Diamagnetic-Electrostatic	Magnet-electric field	Diamagnetic	Trapping	Mix of Surface-micromachining and Bulk	[73]
			Motor		[74, 75]
Diamagnetic-Inductive	Magnet-ac current	Diamagnetic-Conductive material	Gyroscope		[76]
Electrostatic-Inductive	Electric field-ac current	Conductive material	Suspension,	Mix of Surface-micromachining and 3D-micro-coil	
			Transporter		[69]
					[54, 67, 68]
			Accelerator	Surface-micromachining	[70, 77]
			Micro-mirror		[71]

mechanisms, array-based and coherent cooperative actuation. New capabilities such as levitation of arbitrarily shaped micro-objects, performing their linear and angular positioning and transportation within a desired area become feasible. Embedding these new capabilities and mechanisms into micro-systems will build a bridge to the success of today's sensor technology as well. Worth noting that this direction for emerging micro-systems is actively developing in Germany within the framework of the priority programme "Cooperative multilevel multistable micro actuator systems (KOMMMA)" (website of the programme: https://www.spp-kommma. de/index.phpwww.spp-kommma.de) established by German Research Foundation (Deutsche Forschungsgemeinschaft).

References

1. W. Braunbek, Freischwebende Körper im elektrischen und magnetischen Feld. Zeitschrift für Physik **112**(11), 753–763 (1939). https://doi.org/10.1007/BF01339979
2. H. Fujita, Microactuators and micromachines. Proc. IEEE **86**(8), 1721–1732 (1998)
3. D.P. Arnold, N. Wang, Permanent magnets for MEMS. J. Microelectromechanical Syst. **18**(6), 1255–1266 (2009)

4. K. Kratt, V. Badilita, T. Burger, J. Korvink, U. Wallrabe, A fully MEMS-compatible process for 3D high aspect ratio micro coils obtained with an automatic wire bonder. J. Micromechanics Microengineering **20**, 015021 (2010)
5. A.C. Fischer, J.G. Korvink, N. Roxhed, G. Stemme, U. Wallrabe, F. Niklaus, Unconventional applications of wire bonding create opportunities for microsystem integration. J. Micromechanics Microengineering **23**(8), 083001 (2013)
6. S.G. Mariappan, A. Moazenzadeh, U. Wallrabe, Polymer magnetic composite core based microcoils and microtransformers for very high frequency power applications. Micromachines **7**(4), 60 (2016)
7. S. Earnshaw, On the nature of the molecular forces which regulate the constitution of the luminiferous ether. Trans. Camb. Phil. Soc **7**, 97–112 (1842)
8. S. Kumar, D. Cho, W. Carr, A proposal for electrically levitating micromotors. Sensor. Actuat. A-Phys. **24**(2), 141–149 (1990)
9. W. Paul, Electromagnetic traps for charged and neutral particles. Rev. Mod. Phys. **62**, 531–540 (1990). https://doi.org/10.1103/RevModPhys.62.531
10. K.V. Poletkin, A. Asadollahbaik, R. Kampmann, J.G. Korvink, Levitating micro-actuators: a review. Actuators **7**(2), (2018). https://doi.org/10.3390/act7020017
11. S. Kumar, D. Cho, W.N. Carr, Experimental study of electric suspension for microbearings. J. Microelectromechanical Syst. **1**(1), 23–30 (1992). https://doi.org/10.1109/84.128052
12. L. Jin, T. Higuchi, M. Kanemoto, Electrostatic levitator for hard disk media. IEEE Trans. Ind. Electron. **42**(5), 467–473 (1995). https://doi.org/10.1109/41.466330
13. J.U. Jeon, T. Higuchi, Electrostatic suspension of dielectrics. IEEE Trans. Ind. Electron. **45**(6), 938–946 (1998)
14. J. Jin, T.C. Yih, T. Higuchi, J.U. Jeon, Direct electrostatic levitation and propulsion of silicon wafer. IEEE Trans. Ind. Appl. **34**(5), 975–984 (1998). https://doi.org/10.1109/28.720437
15. J.U. Jeon, S.J. Woo, T. Higuchi, Variable-capacitance motors with electrostatic suspension. Sens. Actuators A: Phys. **75**(3), 289–297 (1999). https://doi.org/10.1016/S0924-4247(99)00061-8
16. R.P. Torti, V. Gondhalekar, H. Tran, B. Selfors, S. Bart, B. Maxwell, Electrostatically suspended and sensed micromechanical rate gyroscope, vol. 2220, pp. 2220–2220-12 (1994). https://doi.org/10.1117/12.179613
17. F. Han, Z. Fu, J. Dong, Design and simulation of an active electrostatic bearing for MEMS micromotors, in *2009 4th IEEE International Conference on Nano/Micro Engineered and Molecular Systems* (2009), pp. 80–85. https://doi.org/10.1109/NEMS.2009.5068531
18. F.T. Han, Q.P. Wu, L. Wang, Experimental study of a variable-capacitance micromotor with electrostatic suspension. J. Micromechanics Microengineering **20**(11), 115034 (2010)
19. F.T. Han, L. Wang, Q.P. Wu, Y.F. Liu, Performance of an active electric bearing for rotary micromotors. J. Micromechanics Microengineering **21**(8), 085027 (2011)
20. B. Damrongsak, M. Kraft, A micromachined electrostatically suspended gyroscope with digital force feedback, in *IEEE Sensors, 2005* (2005), pp. 4. https://doi.org/10.1109/ICSENS.2005.1597720
21. B. Damrongsak, M. Kraft, S. Rajgopal, M. Mehregany, Design and fabrication of a micromachined electrostatically suspended gyroscope. Proc. Inst. Mech. Eng. Part C: J. Mech. Eng. Sci. **222**(1), 53–63 (2008). https://doi.org/10.1243/09544062JMES665
22. F.T. Han, Y.F. Liu, L. Wang, G.Y. Ma, Micromachined electrostatically suspended gyroscope with a spinning ring-shaped rotor. J. Micromechanics Microengineering **22**(10), 105032 (2012)
23. B. Sun, F. Han, L. Li, Q. Wu, Rotation control and characterization of high-speed variable-capacitance micromotor supported on electrostatic bearing. IEEE Trans. Ind. Electron. **63**(7), 4336–4345 (2016). https://doi.org/10.1109/TIE.2016.2544252
24. T. Murakoshi, Y. Endo, K. Fukatsu, S. Nakamura, M. Esashi, Electrostatically levitated ring-shaped rotational gyro/accelerometer. Jpn. J. Appl. Phys **42**(4B), 2468–2472 (2003)
25. S. Nakamura, MEMS inertial sensor toward higher accuracy & multi-axis sensing, in *Proc. 4th IEEE Conf. on Sensors* (Irvine, CA 2005), pp. 939–942. https://doi.org/10.1109/ICSENS.2005.1597855

26. R. Toda, N. Takeda, T. Murakoshi, S. Nakamura, M. Esashi, Electrostatically levitated spherical 3-axis accelerometer, in: *Technical Digest. MEMS 2002 IEEE International Conference. Fifteenth IEEE International Conference on Micro Electro Mechanical Systems (Cat. No.02CH37266)* (2002), pp. 710–713. https://doi.org/10.1109/MEMSYS.2002.984369
27. F. Han, Z. Gao, D. Li, Y. Wang, Nonlinear compensation of active electrostatic bearings supporting a spherical rotor. Sens. Actuators A: Phys. **119**(1), 177–186 (2005). https://doi.org/10.1016/j.sna.2004.08.030
28. F. Cui, W. Liu, W. Chen, W. Zhang, X. Wu, Design, fabrication and levitation experiments of a micromachined electrostatically suspended six-axis accelerometer. Sensors **11**(12), 11206–11234 (2011). https://doi.org/10.3390/s111211206
29. F. Han, B. Sun, L. Li, Q. Wu, Performance of a sensitive micromachined accelerometer with an electrostatically suspended proof mass. IEEE Sens. J. **15**(1), 209–217 (2015). https://doi.org/10.1109/JSEN.2014.2340862
30. E.R. Post, G.A. Popescu, N. Gershenfeld, Inertial measurement with trapped particles: a microdynamical system. Appl. Phys. Lett. **96**(14), 143501 (2010). https://doi.org/10.1063/1.3360808
31. W. Rhim, S.K. Chung, D. Barber, K.F. Man, G. Gutt, A. Rulison, R.E. Spjut, An electrostatic levitator for high-temperature containerless materials processing in 1-g. Rev. Sci. Instrum. **64**(10), 2961–2970 (1993). https://doi.org/10.1063/1.1144475
32. X.Y. Ye, Y. Huang, Z.Y. Zhou, Q.C. Li, Q.L. Gong, A magnetic levitation actuator for micro-assembly, in *International Conference on Solid State Sensors and Actuators, 1997. TRANSDUCERS '97*, vol. 2 (IEEE, 1997), pp.797–799. https://doi.org/10.1109/SENSOR.1997.635220
33. C. Ruffert, R. Gehrking, B. Ponick, H.H. Gatzen, Magnetic levitation assisted guide for a linear micro-actuator. IEEE Trans. Magn. **42**(11), 3785–3787 (2006). https://doi.org/10.1109/TMAG.2006.879160
34. C. Dieppedale, B. Desloges, H. Rostaing, J. Delamare, O. Cugat, J. Meunier-Carus, Magnetic bistable micro-actuator with integrated permanent magnets, in *Proceedings of IEEE Sensors, 2004*, vol. 1 (2004), pp. 493–496. https://doi.org/10.1109/ICSENS.2004.1426208
35. C. Ruffert, J. Li, B. Denkena, H.H. Gatzen, Development and evaluation of an active magnetic guide for microsystems with an integrated air gap measurement system. IEEE Trans. Magn. **43**(6), 2716–2718 (2007). https://doi.org/10.1109/TMAG.2007.893779
36. C.R. Dauwalter, J.C. Ha, Magnetically suspended MEMS spinning wheel gyro. IEEE Aerosp. Electron. Syst. Mag. **20**(2), 21–26 (2005). https://doi.org/10.1109/MAES.2005.1397145
37. I.F. Lyuksyutov, D.G. Naugle, K.D.D. Rathnayaka, On-chip manipulation of levitated femto-droplets. Appl. Phys. Lett. **85**(10), 1817–1819 (2004). https://doi.org/10.1063/1.1781735
38. H. Chetouani, C. Jeandey, V. Haguet, H. Rostaing, C. Dieppedale, G. Reyne, Diamagnetic levitation with permanent magnets for contactless guiding and trapping of microdroplets and particles in air and liquids. IEEE Trans. Magn. **42**(10), 3557–3559 (2006). https://doi.org/10.1109/TMAG.2006.880921
39. C. Pigot, H. Chetouani, G. Poulin, G. Reyne, Diamagnetic levitation of solids at microscale. IEEE Trans. Magn. **44**(11), 4521–4524 (2008). https://doi.org/10.1109/TMAG.2008.2003400
40. P. Kauffmann, P. Pham, A. Masse, M. Kustov, T. Honegger, D. Peyrade, V. Haguet, G. Reyne, Contactless dielectrophoretic handling of diamagnetic levitating water droplets in air. IEEE Trans. Magn. **46**(8), 3293–3296 (2010). https://doi.org/10.1109/TMAG.2010.2045361
41. D. Garmire, H. Choo, R. Kant, S. Govindjee, C.H. Sequin, R.S. Muller, J. Demmel, Diamagnetically Levitated MEMS Accelerometers, in, *TRANSDUCERS 2007 - 2007 International Solid-State Sensors, Actuators and Microsystems Conference* (2007), pp. 1203–1206. https://doi.org/10.1109/SENSOR.2007.4300352
42. B. Ando, S. Baglio, V. Marletta, A. Valastro, A short-range inertial sensor exploiting magnetic levitation and an inductive readout strategy. IEEE Trans. Instrum. Meas. **PP**(99), 1–8 (2018). https://doi.org/10.1109/TIM.2017.2785022
43. Y. Su, Z. Xiao, Z. Ye, K. Takahata, Micromachined graphite rotor based on diamagnetic levitation. IEEE Electron Device Lett. **36**(4), 393–395 (2015). https://doi.org/10.1109/LED.2015.2399493

44. Y. Su, K. Zhang, Z. Ye, Z. Xiao, K. Takahata, Exploration of micro-diamagnetic levitation rotor. Jpn. J. Appl. Phys. **56**(12), 126702 (2017). https://doi.org/10.7567/JJAP.56.126702
45. V. Haguet, C. Jeandey, H. Chetouani, G. Reyne, F. Chatelain, Magnetic levitation of micro-droplets in air. Proc. MicroTAS **1**, 110–112 (2006)
46. C. Jeandey, H. Chetouani, V. Haguet, F. Chatelain, G. Reyne, Diamagnetic levitation based digital microfluidics. Proc. MicroTAS **2**, 922–924 (2007)
47. J. Abadie, E. Piat, S. Oster, M. Boukallel, Modeling and experimentation of a passive low frequency nanoforce sensor based on diamagnetic levitation. Sensor. Actuat. A-Phys. **173**(1), 227–237 (2012)
48. X. Chen, A. Keskekler, F. Alijani, P.G. Steeneken, Rigid body dynamics of diamagnetically levitating graphite resonators. Appl. Phys. Lett. **116**(24), 243505 (2020). https://doi.org/10.1063/5.0009604
49. Y. Kim, M. Katsurai, H. Fujita, A proposal for a superconducting actuator using Meissner effect, in *Micro Electro Mechanical Systems, 1989, Proceedings, An Investigation of Micro Structures, Sensors, Actuators, Machines and Robots. IEEE* (1989), pp. 107–112. https://doi.org/10.1109/MEMSYS.1989.77972
50. Y.-K. Kim, M. Katsurai, H. Fujita, A superconducting actuator using the Meissner effect. Sens. Actuators **20**(1), 33–40 (1989). https://doi.org/10.1016/0250-6874(89)87099-4
51. T.A. Coombs, I. Samad, D. Ruiz-Alonso, K. Tadinada, Superconducting micro-bearings. IEEE Trans. Appl. Supercond. **15**(2), 2312–2315 (2005)
52. C. Shearwood, C. Williams, P. Mellor, R. Yates, M. Gibbs, A. Mattingley, Levitation of a micromachined rotor for application in a rotating gyroscope. Electron. Lett. **31**(21), 1845–1846 (1995)
53. N.-C. Tsai, W.-M. Huan, C.-W. Chiang, Magnetic actuator design for single-axis micro-gyroscopes. Microsyst. Technol. **15**, 493–503 (2009)
54. K. Poletkin, Z. Lu, U. Wallrabe, J. Korvink, V. Badilita, Stable dynamics of micro-machined inductive contactless suspensions. Int. J. Mech. Sci. **131–132**, 753–766 (2017). https://doi.org/10.1016/j.ijmecsci.2017.08.016
55. C. Williams, C. Shearwood, P. Mellor, A. Mattingley, M. Gibbs, R. Yates, Initial fabrication of a micro-induction gyroscope. Microelectron. Eng. **30**(1–4), 531–534 (1996)
56. C. Shearwood, C.B. Williams, P.H. Mellor, K.Y. Chang, J. Woodhead, Electro-magnetically levitated micro-discs, in *IEE Colloquium on Microengineering Applications in Optoelectronics* (1996), pp. 6/1–6/3. https://doi.org/10.1049/ic:19960241
57. C. Shearwood, K.Y. Ho, C.B. Williams, H. Gong, Development of a levitated micromotor for application as a gyroscope. Sensor. Actuat. A-Phys. **83**(1–3), 85–92 (2000)
58. V. Badilita, S. Rzesnik, K. Kratt, U. Wallrabe, Characterization of the 2nd generation magnetic microbearing with integrated stabilization for frictionless devices, in **2011 16th International Solid-State Sensors. Actuators and Microsystems Conference** (2011), pp. 1456–1459. https://doi.org/10.1109/TRANSDUCERS.2011.5969624
59. Z. Lu, F. Jia, J. Korvink, U. Wallrabe, V. Badilita, Design optimization of an electromagnetic microlevitation System based on copper wirebonded coils, in *2012 Power MEMS* (Atlanta, GA, USA 2012), pp. 363–366
60. Z. Lu, K. Poletkin, B. den Hartogh, U. Wallrabe, V. Badilita, 3D micro-machined inductive contactless suspension: testing and modeling. Sens. Actuators A: Phys **220**, 134–143 (2014). https://doi.org/10.1016/j.sna.2014.09.017
61. Z. Lu, K. Poletkin, U. Wallrabe, V. Badilita, Performance characterization of micromachined inductive suspensions based on 3D wire-bonded microcoils. Micromachines **5**(4), 1469–1484 (2014). https://doi.org/10.3390/mi5041469
62. K.V. Poletkin, Z. Lu, U. Wallrabe, J.G. Korvink, V. Badilita, A qualitative technique to study stability and dynamics of micro-machined inductive contactless suspensions, in: *2017 19th International Conference on Solid-State Sensors, Actuators and Microsystems (TRANSDUCERS)* (2017), pp. 528–531. https://doi.org/10.1109/TRANSDUCERS.2017.7994102
63. K.V. Poletkin, Z. Lu, A. Moazenzadeh, S.G. Mariappan, J.G. Korvink, U. Wallrabe, V. Badilita, Polymer magnetic composite core boosts performance of three-dimensional micromachined

inductive contactless suspension. IEEE Magn. Lett. **7**, 1–3 (2016). https://doi.org/10.1109/LMAG.2016.2612181

64. K.V. Poletkin, Z. Lu, A. Maozenzadeh, S.G. Mariappan, J. Korvink, U. Wallrabe, V. Badilita, 3D micro-machined inductive suspensions with the lowest energy consumption, in *MikroSystemTechnik Kongress 2017* (VDE Verlag, München, Germany 2017), pp. 500–502

65. K.V. Poletkin, Z. Lu, A. Moazenzadeh, S.G. Mariappan, J.G. Korvink, U. Wallrabe, V. Badilita, Energy aware 3D micro-machined inductive suspensions with polymer magnetic composite core, in *The 17th International Conference on Micro and Nanotechnology for Power Generation and Energy Conversion Applications Power MEMS 2017* (Kanazawa, Japan, 2017), pp. 219–222

66. K.V. Poletkin, A.I. Chernomorsky, C. Shearwood, A proposal for micromachined accelerometer, base on a contactless suspension with zero spring constant. IEEE Sens. J. **12**(07), 2407–2413 (2012). https://doi.org/10.1109/JSEN.2012.2188831

67. K.V. Poletkin, A novel hybrid contactless suspension with adjustable spring constant, in *2017 19th International Conference on Solid-State Sensors, Actuators and Microsystems (TRANSDUCERS)* (2017), pp. 934–937. https://doi.org/10.1109/TRANSDUCERS.2017.7994203

68. K.V. Poletkin, J.G. Korvink, Modeling a pull-in instability in micro-machined hybrid contactless suspension. Actuators **7**(1), 11 (2018). https://doi.org/10.3390/act7010011

69. K. Poletkin, Z. Lu, U. Wallrabe, V. Badilita, A new hybrid micromachined contactless suspension with linear and angular positioning and adjustable dynamics. J. Microelectromechanical Syst. **24**(5), 1248–1250 (2015). https://doi.org/10.1109/JMEMS.2015.2469211

70. I. Sari, M. Kraft, A MEMS linear accelerator for levitated micro-objects. Sensor. Actuat. A-Phys. **222**, 15–23 (2015)

71. Q. Xiao, Y. Wang, S. Dricot, M. Kraft, Design and experiment of an electromagnetic levitation system for a micro mirror. Microsyst. Technol. **25**(8), 3119–3128 (2019)

72. W. Zhang, W. Chen, X. Zhao, X. Wu, W. Liu, X. Huang, S. Shao, The study of an electromagnetic levitating micromotor for application in a rotating gyroscope. Sensor. Actuat. A-Phys. **132**(2), 651–657 (2006)

73. P. Kauffmann, J. Nussbaumer, A. Masse, C. Jeandey, H. Grateau, P. Pham, G. Reyne, V. Haguet, Self-arraying of charged levitating droplets. Anal. Chem. **83**(11), 4126–4131 (2011). https://doi.org/10.1021/ac2002774

74. W. Liu, W.-Y. Chen, W.-P. Zhang, X.-G. Huang, Z.-R. Zhang, Variable-capacitance micromotor with levitated diamagnetic rotor. Electron. Lett. **44**(11), 681–683 (2008). https://doi.org/10.1049/el:20080528

75. Y. Xu, Q. Cui, R. Kan, H. Bleuler, J. Zhou, Realization of a diamagnetically levitating rotor driven by electrostatic field. IEEE/ASME Trans. Mechatron. **22**(5), 2387–2391 (2017). https://doi.org/10.1109/TMECH.2017.2718102

76. K. Liu, W. Zhang, W. Liu, W. Chen, K. Li, F. Cui, S. Li, An innovative micro-diamagnetic levitation system with coils applied in micro-gyroscope. Microsyst. Technol. **16**(3), 431 (2009). https://doi.org/10.1007/s00542-009-0935-x

77. I. Sari, M. Kraft, A micro electrostatic linear accelerator based on electromagnetic levitation, in *2011 16th International Solid-State Sensors. Actuators and Microsystems Conference* (2011), pp. 1729–1732. https://doi.org/10.1109/TRANSDUCERS.2011.5969625

Chapter 2
Micro-Coil Fabrication Techniques

Inductive levitation micro-systems have been actively studied since 1995 when the
Shearwood group [1] fabricated the first prototype with targeted applications including micro-motors, -actuators and -inertial sensors. Then, in 2006, the Zhang group
presented the IL-micro-system with improved planar coil design [2]. An alternative
coil design in the shape of a rectangular spiral, which also provides stable levitation,
was employed in the micro-gyroscope prototype reported by the Tsai group in work
[3]. All the aforementioned research groups used surface micromachining technology
for coil fabrication. Due to this technology, the coils and levitated object dissipated
a lot of energy. As a result, the observed temperature of coils was higher 150 °C.
In 2011, the Wallrabe and Korvink group developed jointly 3D coil technology [4].
Application of this emerging technology has reduced dramatically the energy dissipation in IL-micro systems. These two techniques for micro-coil fabrication process
are discussed in the proceeding chapter.

2.1 Planar Coil Technology

In general, the micromachined fabrication process of micro-coils focuses on the
following target characteristics:

- high electrical quality factor and hence low electrical resistance;
- an increase of number of winding turns to reduce the amplitude of the coil current;
- electrical winding insulation;
- high mechanical strength and stability;
- low manufacturing and assembly costs.

© The Editor(s) (if applicable) and The Author(s), under exclusive license   17
to Springer Nature Switzerland AG 2021
K. Poletkin, *Levitation Micro-Systems*, Microsystems and Nanosystems,
https://doi.org/10.1007/978-3-030-58908-0_2

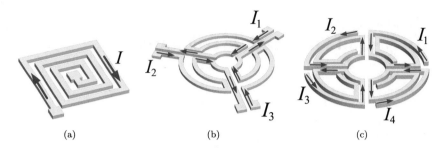

Fig. 2.1 Designs of planar micro-coils: I is the electric current

The designs of the *planar micro-coils* applied in IL-micro-systems are shown in
Fig. 2.1. Coil design shown in Fig. 2.1a provides only stable levitation of a levitated
micro-object, which can have a disc or square shape. Figure 2.1b, c demonstrate
designs, which in addition to stable levitation allow us to rotate the disc-shaped
micro-object. The rotation is reached by using poles, which are sequentially excited
by the high-frequency current. The sequential excitation of poles by ac currents
having particular phase-shifts to each other (depending on a number of poles) rotates
the magnetic flux generated by coils, which in turn involves a levitated micro-disc
in the revolution. Actually, the rotation principle corresponds to an asynchronous
electromagnetic motor.

Planar micro-coil fabrication exploits surface micromachining technology, which
includes the following essential techniques:

- metal evaporation;
- electroplating;
- wet chemical etching.

The outline of the fabrication process for one-layer planar micro-coil is shown in
Fig. 2.2. As seen from Fig. 2.2 the fabrication of one-layer coil is relatively simple.
As a result within the framework of one layer, designing of the coil becomes flexible
allowing for the design of a complex coil structure. However, the fabrication complexity is dramatically increased once the number of layers and consequently the
number of coil turns is increased.

2.2   3D Micro-Coil Technology

Using *3D micro-coil* technology, a micro-coil with an unlimited number of turns
can be fabricated. Moreover, irrespective of the number of turns, only two masks
are required for the fabrication. The typical design of 3D coils, which is applied
to IL-micro-systems, is shown in Fig. 2.3a. This design includes two 3D micro-
coils, namely, stabilization and levitation ones. As an example, Fig. 2.3a shows the
fabricated 3D micro-coil structure having following dimensions: a radius of the

Process:

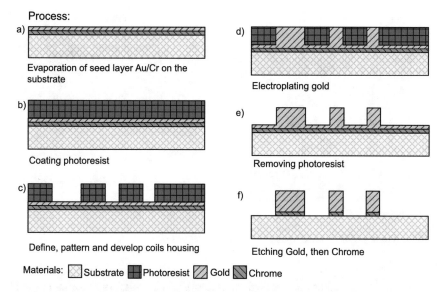

Fig. 2.2 Outline of the fabrication process for one layer planar micro-coil

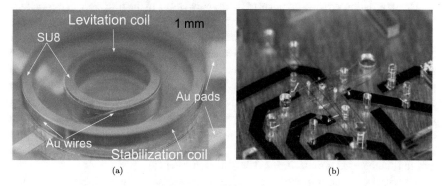

Fig. 2.3 **a** The typical fabricated 3D micro-coil structure for IL-micro-systems (this is a link on YouTube video: https://youtu.be/uqind0ATEIE); **b** Wirebonded nonsolenoidal structure realized in cooperation of Laboratory for Microactuators and Laboratory for Simulation

levitation coil is 1000 μm, a radius of the stabilization coil is 1900 μm, a height of both pillars is 700 μm, number of windings for stabilization and levitation coil are to be 12 and 20, respectively, and a diameter of gold wire is 25 μm. Note that to avoid throwing the wire out the post, the top turn of each winding has a distance from the top of the pillar in a range 30–100 μm. In the presented design, this distance is to be 50 μm.

Also, 3D coil technology allows fabricating none cylindrical-shaped coils. As an example of the flexibility of the 3D coil fabrication technique, a 3D structure base on non-cylindrical-shaped micro-coils is shown in Fig. 2.3b, which was fabricated

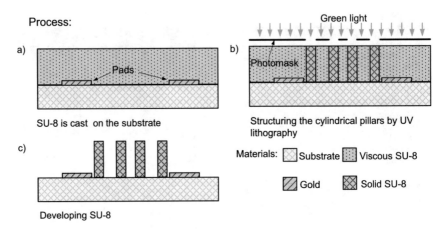

Fig. 2.4 Outline of the fabrication process of the cylindrical pillars

in cooperation of Laboratory for Microactuators and Laboratory for Simulation at IMTEK, Freiburg University, and proposed to be used as gradient coil [5].

The fabrication process consists of three main steps. In the first step, using surface micromachining, pads for electrical contacts are fabricated on the substrate by the same process outlined in Fig. 2.2, where the first mask is used. In the second step, usually 600–700 μm-thick SU-8 2150 is cast on the substrate. Using the second mask, the cylindrical pillars are structured by UV lithography. An outline of this process is shown in Fig. 2.4. In the last step, the coils are manufactured using an automatic *wire-bonder* which allows us to freely define the total number of windings per coil, the pitch between the windings, and the number of winding layers. Although using a wire-bonder for the coil fabrication is a serial process, this process is very fast—hundreds of milliseconds per coil depending on the exact dimensions—and perfectly integrated with traditional MEMS processes [4].

3D micro-coils opposed to planar micro-coils dramatically increases the number of winding turns and at the same time drastically reduces the fabrication complexity. By embracing 3D micro-coils, energy dissipation is significantly reduced. Even more, the operating temperature of 3D coils can be further decreased to the ambient temperature by integrating them with polymer magnetic composite core as it is discussed in Sect. 5.2. The above-mentioned advantages of the 3D micro-coils, which is supported by using 3D coil technology, improve further the IL-micro-system performance and open a new application, for instance, in micro-transporters, micro-actuators, micro-accelerators and micro-inertial sensors as considered below.

References

1. C. Shearwood, C. Williams, P. Mellor, R. Yates, M. Gibbs, A. Mattingley, Levitation of a micromachined rotor for application in a rotating gyroscope. Electron. Lett. **31**(21), 1845–1846 (1995)
2. W. Zhang, W. Chen, X. Zhao, X. Wu, W. Liu, X. Huang, S. Shao, The study of an electromagnetic levitating micromotor for application in a rotating gyroscope. Sensor. Actuat. A-Phys. **132**(2), 651–657 (2006)
3. N.-C. Tsai, W.-M. Huan, C.-W. Chiang, Magnetic actuator design for single-axis micro-gyroscopes. Microsyst Technol **15**, 493–503 (2009)
4. K. Kratt, V. Badilita, T. Burger, J. Korvink, U. Wallrabe, A fully MEMS-compatible process for 3D high aspect ratio micro coils obtained with an automatic wire bonder. J. Micromechanics Microengineering **20**, 015021 (2010)
5. A.C. Fischer, J.G. Korvink, N. Roxhed, G. Stemme, U. Wallrabe, F. Niklaus, Unconventional applications of wire bonding create opportunities for microsystem integration. J. Micromechanics Microengineering **23**(8), 083001 (2013)

Chapter 3
Analytical Modelling

To fully benefit from inductive levitation and utilize its advantages in L-micro-systems such as

- passive levitation;
- established the micro-fabrication process;
- proved energy efficiency at micro-level,

adequate and accurate mathematical models based on differential equations describing diverse physical processes and electromagnetic phenomena, as well as their combinations in the designed IL-micro-systems, are required to be developed. A mathematical model assumes to apply a certain idealization, which can never be avoided for a system to simplify its true physical properties. Indeed, an enormous spectrum of physical processes occurring in micro-systems upon applying such an idealization can be reduced to a problem, in which the spatial configuration of the system is completely defined by a particular limited number of *generalized coordinates* and *generalized velocities*. In this case, differential equations can be compiled by *analytical methods*, in particular, including methods developed in analytical mechanics based on *Lagrangian* and *Hamiltonian* formalism.

The stability of L-micro-systems is the main and crucial issue requiring to be addressed at the beginning of designing. To solve it, all four types of forces, namely, *dissipation*, *gyroscopic*, *potential* and *circulatory* ones and concomitant nonlinear effects due to pull-in phenomenon, mechanical contacts and impacts must be taken into account. The description of electromagnetic levitation requires the application of the Maxwell equations. Although these equations are universally applicable, their application even for simple designs is an extremely difficult task [1]. Even these designs are studied numerically by using commercially available software, this task is still a challenge [2–5], which is not able to cover all aspects including stability and nonlinear dynamical response of the system [6].

To avoid having a deal with field equations, analytical and quasi-finite element approaches for modelling IL-micro-systems based on the Lagrangian formalism are developed. The mathematical formulation for both approaches becomes possible due

© The Editor(s) (if applicable) and The Author(s), under exclusive license
to Springer Nature Switzerland AG 2021
K. Poletkin, *Levitation Micro-Systems*, Microsystems and Nanosystems,
https://doi.org/10.1007/978-3-030-58908-0_3

to the recent progress in analytical calculation of mutual inductance between filament loops [7, 8] and the analytical technique proposed by the author in work [9] and generalized in paper [10], in which the induced eddy current into a levitated micro-object is approximated by a finite collection of eddy current circuits. For instance, as the advantage of the developed analytical model can be considered in fact that it allows us to investigate the general stability properties of an IL-micro-system as a dynamic system, and these properties are synthesized in three major theorems. In particular, it was proved that the stable levitation in the IL-micro-system without damping is impossible. Based on the developed approach, general guidelines for designing IL-micro-systems are provided. In addition to the successful application, this technique to study the dynamics and stability of the symmetric and axially symmetric designs of IL-micro-systems both based on 3D micro-coil technology is demonstrated. Also, the model allows predicting the static and dynamic pull-in parameters of HL-micro-systems without needs for solving nonlinear differential equations.

3.1 Analytical Mechanics of Micro-Electro-Mechanical Systems

The systems with essentially coupling mechanical and electromagnetic processes, which occur on a micrometre scale, are referred to micro-electro-mechanical systems (MEMS). Studying such systems requires at the beginning to formulate the prime assumptions and to accept models, which account for only an essential phenomenon (the major physical properties) and neglect other phenomena of minor influence [11].

It is assumed, as usual in analytical mechanics, that the state of a mechanical part of MEMS can be modelled by n independent *generalized coordinates* $\underline{q} = [q_1 \ \ldots \ q_n]^T$. The number of these independent coordinates is defined by constraints applied to the system, which are assumed to be holonomic and ideal [12]. For instance, let us consider a system of N point masses, which represents the mechanical part of MEMS, and this system has s holonomic constraints, then the number n of independent generalized coordinates is qual to $3N - s$. In a case of a rigid body, the number of coordinates is reduced to $6 - s$ and, if no constraints ($s = 0$), then is equal just to six.

In order to define the state of an electromagnetic part of MEMS, the spatial distribution of vectors of the magnetic flux density B and the electric field intensity E, as well their changing with time must be known. Considering the system as continuous one, the description of its state requires the application of the infinite number of time functions. However, vectors of B and E can be also modelled by means of electrical and magnetic circuits utilizing a finite number of parameters such as electrical current, voltage, magnetic flux, etc. Replacing vectors of magnetic and electric fields through these parameters is equivalent to introducing the generalized coordinates into the electromagnetic system. Let us assume that the electromagnetic part of MEMS is represented by m electric circuits, each of them consists of the capacitor, inductor and resistor connecting in series. These electric circuits have no direct electrical connection to each of other, but of course, they are coupled electromagnetically. Such a representation becomes possible if conditions of *quasistationarity* are held. Also, it is assumed that a cross-section of conductor is much smaller than its length. Then, for alternating current with a relatively high frequency, the inductance of circuit is defined by a shape of the conductor. The effect of the distribution of the current along the cross-section of the conductor on the inductance of the circuit can be neglected.

Let us denote i_k ($k = 1, 2, \ldots, m$) as the current in the k-circuit and u_k as the electromotive force (e.m.f) applied to the k-circuit. The charges accumulated by capacitors are denoted as e_k ($k = 1, 2, \ldots, m$) and they have the following relation with the current $de_k = i_k dt$. Then, the resistor and capacitor belonged to the k-circuit are denoted as R_k and C_k, respectively. In general, the relative motion of mechanical elements of MEMS leads to changing in positions of electrodes of capacitors with respect to each other and the shape of conductors. Consequently, it is assumed that the capacitor and inductor are functions of generalized coordinates, namely, $C_k = C_k(q_1, \ldots, q_n)$, $L_{kk} = L_{kk}(q_1, \ldots, q_n)$ and $L_{kr} = L_{kr}(q_1, \ldots, q_n)$, where L_{kk} is the self-inductance of the k-circuit and L_{kr} ($k \neq r$) is the mutual inductance between the k- and r-circuit.

Now, the Lagrange function of the electromechanical system can be compiled in the following form:

$$L = K(q_j, \dot{q}_j) - \Pi(q_j) + W_m(q_j, i_k) - W_e(q_j, e_k), \tag{3.1}$$

where $K(q_j, \dot{q}_j)$ and $\Pi(q_j)$ are the kinetic and potential energy of the system, respectively, $W_m(q_j, i_k)$ and $W_e(q_j, e_k)$ are magnetic and electrical energy stored in the system, respectively. In a case of stationary constraints, the kinetic energy of the system has the quadratic form such as

$$K = \frac{1}{2} \sum_i^n \sum_j^n a_{ij} \dot{q}_i \dot{q}_j, \tag{3.2}$$

where $a_{ij} = a_{ij}(q_1, \ldots, q_n, t)$ are coefficients of inertia, depending on the generalized coordinates. The potential energy Π is also the function of the generalized coordinates, namely,

$$\Pi = \Pi(q_1, \ldots, q_n). \tag{3.3}$$

The electrical energy stored between the electrodes of capacitors can be found by the following known equation:

$$W_e = \frac{1}{2} \sum_k^m \frac{e_k^2}{C_k}. \tag{3.4}$$

The magnetic energy stored by the system of m-loops with electrical currents is

$$W_m = \frac{1}{2} \sum_{k}^{m} \sum_{r}^{m} L_{kr} i_k i_r. \tag{3.5}$$

Besides, the energy from the system is dissipated due to Joule heating and friction. This dissipation energy can be taken into account by introducing the following function, which sums up the dissipation occurring in electrical and mechanical parts of the system, such as

$$\Psi = \Psi_e(i_k) + \Psi_{mech}(q_j, \dot{q}_j). \tag{3.6}$$

Noting that the electrical dissipation function Ψ_e is independent of the generalized coordinates and the mechanical dissipation function Ψ_{mech} of the currents, thus it can be concluded that

$$\frac{\partial \Psi}{\partial i_k} = \frac{\partial \Psi_e}{\partial i_k}; \quad \frac{\partial \Psi}{\partial \dot{q}_j} = \frac{\partial \Psi_{mech}}{\partial \dot{q}_j}. \tag{3.7}$$

Using lagrangian (3.1) and dissipation function (3.6), the set of the Lagrange-Maxwell equations of electromechanical system can be obtained:

$$\begin{cases} \dfrac{d}{dt}\left(\dfrac{\partial L}{\partial \dot{e}_k}\right) - \dfrac{\partial L}{\partial e_k} + \dfrac{\partial \Psi}{\partial \dot{e}_k} = u_k; \ k = 1, \ldots, m; \\[2ex] \dfrac{d}{dt}\left(\dfrac{\partial L}{\partial \dot{q}_j}\right) - \dfrac{\partial L}{\partial q_j} + \dfrac{\partial \Psi}{\partial \dot{q}_j} = Q_j; \ j = 1, \ldots, n, \end{cases} \tag{3.8}$$

where Q_j is the generalized forces applied to the system. The obtained equations are called the Lagrange-Maxwell equations. Equations (3.8) is the set of $n + m$ ordinary differential equations of the second order with respect to the generalized coordinates and charges. Worth noting that methods of analytical mechanics were developed by two physicists, namely, Joseph Louis Lagrange (1736–1813) and William Rowan Hamilton (1805–1865) in eighteenth and nineteenth centuries. In particular, in 1873 using the Lagrange approach, James Clerk Maxwell (1831–1879) derived the equations describing electromechanical systems in his famous work "A Treatise on Electricity and Magnetism" [13].

In general, set (3.8) is nonlinear because of the terms, which are defined by

$$\sum_{j}^{n} \sum_{k}^{m} \frac{\partial L_{kr}}{\partial q_j} \dot{q}_j i_k, \tag{3.9}$$

representing the induced e.m.f by moving conductors with currents. Set (3.8) includes also ponderomotive forces, which are

$$Q_j^* = \frac{\partial}{\partial q_j}(W_m - W_e) = \frac{1}{2} \sum_{k}^{m} \sum_{r}^{m} \frac{\partial L_{kr}}{\partial q_j} i_k i_r + \frac{1}{2} \sum_{k}^{m} \frac{\partial C_k}{\partial q_j} \frac{e_k^2}{C_k^2}. \tag{3.10}$$

Formally, ponderomotive forces arise as a result of changing the electrical and magnetic energies with the generalized coordinates. The generalized coordinates q_j determine the positions of electrodes of capacitors and wires of conductors, as well as the
distribution of magnetic permeability μ and dielectric constant ε in space. For this,
the different types of ponderomotive forces are responsible for the force interaction,
for instance, the force interaction between wires with currents, the force interaction
between charged bodies, force acting on a body due to the difference between the
magnetic permeability of the body and surrounding space and etc. However, in any
case, equation for ponderomotive force (3.10) is still applicable.

For some treatments, it is convenient to define the state of the electromechanical
system through *the generalized momentum*, which is defined as follows:

$$p_j = \frac{\partial L}{\partial \dot{q}_j}, \tag{3.11}$$

and *the generalized magnetic flux* such as

$$\phi_k = \frac{\partial L}{\partial \dot{e}_k}. \tag{3.12}$$

Using the generalized momentum and magnetic flux defined by (3.11) and (3.12),
respectively, the *hamiltonian* of the system can be explicitly obtained through the
Legendre transform of the *lagrangian* such as

$$H(p, \phi, q, e) = p \cdot \dot{q} + \phi \cdot i - L(\dot{q}, q, i, e). \tag{3.13}$$

Then $n + m$ Lagrange-Maxwell equations can be equivalently represented by the set
of $2(n + m)$ first-order Hamilton's equations as follows:

$$\begin{cases} \dfrac{d}{dt} e_k = \dfrac{\partial H}{\partial \phi_k}; \ k = 1, \ldots, m; \\ \dfrac{d}{dt} \phi_k = u_k - \dfrac{\partial H}{\partial e_k}; \\ \dfrac{d}{dt} q_j = \dfrac{\partial H}{\partial p_j}; \ j = 1, \ldots, n; \\ \dfrac{d}{dt} p_j = Q_j - \dfrac{\partial H}{\partial q_j}. \end{cases} \tag{3.14}$$

3.2 Statement of the Problem for Modelling

Let us consider the schematic of an inductive levitation system shown in Fig. 3.1,
which consists of a system of N coils, and a levitated micro-object. Each j-th coil
is fed by its own alternating current denoted by i_{cj} and generates a time-variable

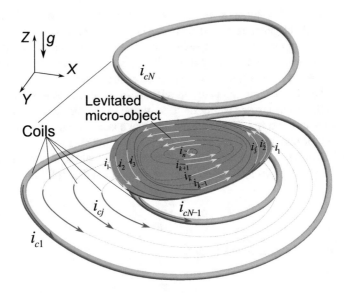

Fig. 3.1 General schematic of levitation system with the n eddy current circuits: YXZ is the fixed coordinate frame; g is the gravity acceleration directed along the Z axis; i_k is the eddy current in the k-th eddy current circuit; i_{cj} is the current in the j-coil

magnetic field in space. In turn, the alternating magnetic flux passing through the conducting micro-object induces an eddy current. The eddy current is continuously distributed within the micro-object, however, this distribution is not homogeneous. This fact helps us to selectively choose the n eddy current circuits having the representative behaviour of the entire eddy current distribution as shown in Fig. 3.1. As seen from Fig. 3.1, i_k is the eddy current in the k-th eddy current circuit. The interaction between the currents in the coils and the eddy current produces the repulsive force levitating the micro-object at an equilibrium position, which can be characterized with respect to the fixed coordinate frame YXZ. Considering this micro-object as a rigid body, its behaviour relative to the equilibrium position can be characterized in general by six generalized coordinates corresponding to three linear and three angular displacements. Let us denote generalized coordinates of linear displacements as q_l ($l = 1, 2, 3$), while of angular displacements as q_l ($l = 4, 5, 6$). It is convenient to use the Bryan angles (Cardan angles) to characterize the angular position of a levitated micro-object.

Adapting the generalized coordinates and the assumptions introduced above, the model can be written by using the Lagrange- Maxwell equations as follows:

$$\begin{cases} \dfrac{d}{dt}\left(\dfrac{\partial L}{\partial i_k}\right) + \dfrac{\partial \Psi}{\partial i_k} = 0; \ k = 1, \ldots, n; \\[2mm] \dfrac{d}{dt}\left(\dfrac{\partial L}{\partial \dot{q}_l}\right) - \dfrac{\partial L}{\partial q_l} + \dfrac{\partial \Psi}{\partial \dot{q}_l} = F_l; \ l = 1, 2, 3; \\[2mm] \dfrac{d}{dt}\left(\dfrac{\partial L}{\partial \dot{q}_l}\right) - \dfrac{\partial L}{\partial q_l} + \dfrac{\partial \Psi}{\partial \dot{q}_l} = T_l; \ l = 4, 5, 6, \end{cases} \tag{3.15}$$

where $L = K - \Pi + W_m$ is the Lagrange function for the micro-object-coil system; $K = K(\dot{q}_1, \ldots, \dot{q}_6, q_4, \ldots, q_6)$ is the kinetic energy of the system; $\Pi = \Pi(q_1, \ldots, q_6)$ is the potential energy of the system; $W_m = W_m(q_1, \ldots, q_6, i_{c1}, \ldots, i_{cN}, i_1, \ldots, i_n)$ is the energy stored in the electromagnetic field; $\Psi = \Psi(\dot{q}_1, \ldots, \dot{q}_6, i_1, \ldots, i_n)$ is the dissipation function; F_l and T_l ($l = 1, 2, 3$) are the generalized forces and torques, respectively, acting on the micro-object relative to the appropriate generalized coordinates.

The kinetic energy (for small displacements) is

$$T = \frac{1}{2} \sum_{l=1}^{3} m\dot{q}_l^2 + \frac{1}{2} \sum_{l=1}^{3} J_l \dot{\omega}_l^2, \tag{3.16}$$

where m is the mass of the micro-object; J_l is its moment of inertia in terms of the appropriate generalized angular coordinates, $\omega_l = \omega_l(q_4, \ldots, q_6, \dot{q}_4, \ldots, \dot{q}_6)$ ($l = 1, 2, 3$) are the components of the vector $\boldsymbol{\omega}$ of angular velocity of the micro-object relative to the fixed coordinate frame.

The linear generalized coordinates, q_l, $l = 1, 2, 3$, are defined in the orthogonal coordinate frame. Hence, for the further simplification of analysis, it can be assumed that the generalized coordinate q_3 is directed along the gravity acceleration g, as shown in Fig. 3.1. Then, the potential energy can be defined simply as follows:

$$\Pi = mgq_3. \tag{3.17}$$

The dissipation function is

$$\Psi = \frac{1}{2} \sum_{r=1}^{6} \mu_r \dot{q}_r^2 + \frac{1}{2} \sum_{k=1}^{n} R_k i_k^2 \pm \sum_{k=1}^{n} \sum_{s=1, s \neq k}^{n} R_{ks} i_k i_s, \tag{3.18}$$

where μ_r is the damping coefficient corresponding to the appropriate generalized coordinates; R_k is the electrical resistance for the k-th eddy current circuit within the micro-object; R_{ks} is the resistance of a common circuit for k-th and s-th eddy current circuits (e.g., this case is shown in Fig. 7.1 for eddy currents i_1, i_2 and i_3). For generality, it is assumed that the k-th eddy current may share a common path with the s-th eddy current. The plus-minus sign corresponds to eddy currents having the same or opposite direction on the common circuit. The energy stored within the electromagnetic field can be written as

$$W_m = \frac{1}{2} \sum_{j=1}^{N} \sum_{s=1}^{N} L_{js}^c i_{cj} i_{cs} + \frac{1}{2} \sum_{k=1}^{n} \sum_{s=1}^{n} L_{ks}^o i_k i_s + \sum_{k=1}^{n} \sum_{j=1}^{N} M_{kj} i_k i_{cj}, \qquad (3.19)$$

where L_{jj}^c is the self inductance of the j-coil; L_{js}^c, $j \neq s$ is the mutual inductance between j- and s-coils; $L_{kk}^o = L_{kk}^o(q_1, \ldots, q_6)$ is the self inductance of the k-eddy current circuit; $L_{ks}^o = L_{ks}^o(q_1, \ldots, q_6)$, $k \neq s$ is the mutual inductance between k- and s-eddy current circuits; $M_{kj} = M_{kj}(q_1, \ldots, q_6)$ is the mutual inductance between the k-eddy current circuit and the j-coil.

We now show that the induced eddy currents i_k can be expressed in terms of coil currents i_{cj} under a particular condition discussed below. Assuming that for each coil, the current i_{cj} is a periodic signal with an amplitude of I_{cj} at the same frequency f, we can write

$$i_{cj} = I_{cj} e^{Jft}, \qquad (3.20)$$

where $J = \sqrt{-1}$. Taking the derivative of the Lagrange function with respect to the eddy current i_k, we have

$$\frac{\partial L}{\partial i_k} = \frac{\partial W_m}{\partial i_k} = \sum_{s=1}^{n} L_{ks}^o i_s + \sum_{j=1}^{N} M_{kj} i_{cj}, \qquad (3.21)$$

or

$$\frac{\partial L}{\partial i_k} = \frac{\partial W_m}{\partial i_k} = L_{kk}^o i_k + \sum_{s=1, s \neq k}^{n} L_{ks}^o i_s + \sum_{j=1}^{N} M_{kj} i_{cj}. \qquad (3.22)$$

Substituting (3.22) into (3.15), the first equation of set (3.15) becomes

$$\frac{d}{dt}\left(\frac{\partial L}{\partial i_k}\right) + \frac{\partial \Psi}{\partial i_k} = \sum_{r=1}^{6} \frac{\partial L_{kk}^o}{\partial q_r} \frac{dq_r}{dt} i_k + L_{kk}^o \frac{di_k}{dt}$$

$$+ \sum_{s=1, s \neq k}^{n} \left(\sum_{r=1}^{6} \frac{\partial L_{ks}^o}{\partial q_r} \frac{dq_r}{dt} i_s + L_{ks}^o \frac{di_s}{dt} \right) \qquad (3.23)$$

$$+ \sum_{j=1}^{N} \left(\sum_{r=1}^{6} \frac{\partial M_{kj}}{\partial q_r} \frac{dq_r}{dt} i_{cj} + M_{kj} \frac{di_{cj}}{dt} \right) + R_k i_k \pm \sum_{s=1, s \neq k}^{n} R_{ks} i_s = 0.$$

Accounting for (3.20), the k- eddy current can be represented as

$$i_k = I_k e^{Jft}, \qquad (3.24)$$

where I_k is the amplitude. Hence, Eq. (3.23) can be rewritten in term of the current amplitudes as follows:

$$\sum_{r=1}^{6} \frac{\partial L_{kk}^{o}}{\partial q_r} \frac{dq_r}{dt} I_k + L_{kk}^{o} Jf I_k$$

$$+ \sum_{s=1,\, s\neq k}^{n} \left(\sum_{r=1}^{6} \frac{\partial L_{ks}^{o}}{\partial q_r} \frac{dq_r}{dt} I_s + L_{ks}^{o} Jf I_s \right) \tag{3.25}$$

$$+ \sum_{j=1}^{N} \left(\sum_{r=1}^{6} \frac{\partial M_{kj}}{\partial q_r} \frac{dq_r}{dt} I_{cj} + M_{kj} Jf I_{cj} \right) + R_k I_k \pm \sum_{s=1,\, s\neq k}^{n} R_{ks} I_s = 0.$$

Equation (3.25) is nonlinear due to the velocities of generalized coordinates, dq_r/dt. In fact, the analysis of the existing levitation micro-system prototypes shows that the velocity dq_r/dt can be assumed to be small. Also the frequency f is usually larger than 1 MHz, which corresponds to $\sim10^7$ rad/s. Hence, Eq. (3.25) can be rewritten [14] as follows:

$$\left(L_{kk}^{o} + R_k/(Jf) \right) I_k + \sum_{s=1,\, s\neq k}^{n} (L_{ks}^{o} \pm R_{ks}/(Jf)) I_s = -\sum_{j=1}^{N} M_{kj} I_{cj}. \tag{3.26}$$

It is important to note that, for higher values of the generalized velocities dq_r/dt, when the quasi-static approximation does not hold, Eq. (3.25) must be used. In order to define the eddy currents I_k, a set of linear equations can be compiled from (3.26) in a matrix form as follows:

$$\begin{bmatrix} L_{11}^{o}+\frac{R_1}{Jf} & L_{12}^{o}\pm\frac{R_{12}}{Jf} & \cdots & L_{1k}^{o}\pm\frac{R_{1k}}{Jf} & \cdots & L_{1n}^{o}\pm\frac{R_{1n}}{Jf} \\ L_{21}^{o}\pm\frac{R_{21}}{Jf} & L_{22}^{o}+\frac{R_2}{Jf} & \cdots & L_{2k}^{o}\pm\frac{R_{2k}}{Jf} & \cdots & L_{2n}^{o}\pm\frac{R_{2n}}{Jf} \\ \vdots & \vdots & \ddots & \vdots & \ddots & \vdots \\ L_{k1}^{o}\pm\frac{R_{k1}}{Jf} & L_{k2}^{o}\pm\frac{R_{k2}}{Jf} & \cdots & L_{kk}^{o}+\frac{R_k}{Jf} & \cdots & L_{kn}^{o}\pm\frac{R_{kn}}{Jf} \\ \vdots & \vdots & \ddots & \vdots & \ddots & \vdots \\ L_{n1}^{o}\pm\frac{R_{n1}}{Jf} & L_{n2}^{o}\pm\frac{R_{n2}}{Jf} & \cdots & L_{nk}^{o}\pm\frac{R_{nk}}{Jf} & \cdots & L_{nn}^{o}+\frac{R_{nn}}{Jf} \end{bmatrix} \begin{bmatrix} I_1 \\ I_2 \\ \vdots \\ I_k \\ \vdots \\ I_n \end{bmatrix} = \begin{bmatrix} -\sum_{j=1}^{N} M_{1j} I_{cj} \\ -\sum_{j=1}^{N} M_{2j} I_{cj} \\ \vdots \\ -\sum_{j=1}^{N} M_{kj} I_{cj} \\ \vdots \\ -\sum_{j=1}^{N} M_{nj} I_{cj}, \end{bmatrix}, \tag{3.27}$$

where $L_{ks}^{o} = L_{sk}^{o}$. The solution of (3.27) for I_k can be found by using Cramer's rule and is written as follows:

$$I_k = \frac{\Delta_k}{\Delta}, \tag{3.28}$$

where

$$\Delta = \begin{vmatrix} L_{11}^{o}+\frac{R_1}{Jf} & L_{12}^{o}\pm\frac{R_{12}}{Jf} & \cdots & L_{1k}^{o}\pm\frac{R_{1k}}{Jf} & \cdots & L_{1m}^{o}\pm\frac{R_{1m}}{Jf} \\ L_{21}^{o}\pm\frac{R_{21}}{Jf} & L_{22}^{o}+\frac{R_2}{Jf} & \cdots & L_{2k}^{o}\pm\frac{R_{2k}}{Jf} & \cdots & L_{2m}^{o}\pm\frac{R_{2m}}{Jf} \\ \vdots & \vdots & \ddots & \vdots & \ddots & \vdots \\ L_{k1}^{o}\pm\frac{R_{k1}}{Jf} & L_{k2}^{o}\pm\frac{R_{k2}}{Jf} & \cdots & L_{kk}^{o}+\frac{R_k}{Jf} & \cdots & L_{km}^{o}\pm\frac{R_{km}}{Jf} \\ \vdots & \vdots & \ddots & \vdots & \ddots & \vdots \\ L_{m1}^{o}\pm\frac{R_{m1}}{Jf} & L_{m2}^{o}\pm\frac{R_{m2}}{Jf} & \cdots & L_{mk}^{o}\pm\frac{R_{mk}}{Jf} & \cdots & L_{mm}^{o}+\frac{R_m}{Jf} \end{vmatrix}, \tag{3.29}$$

$$
\Delta_k =
\begin{vmatrix}
L_{11}^o + \frac{R_1}{jf} & L_{12}^o \pm \frac{R_{12}}{jf} & \cdots & -\sum_{j=1}^{n} M_{1j} I_{cj} & \cdots & L_{1m}^o \pm \frac{R_{1m}}{jf} \\
L_{21}^o \pm \frac{R_{21}}{jf} & L_{22}^o + \frac{R_2}{jf} & \cdots & -\sum_{j=1}^{n} M_{2j} I_{cj} & \cdots & L_{2m}^o \pm \frac{R_{2m}}{jf} \\
\vdots & \vdots & \ddots & \vdots & \ddots & \vdots \\
L_{k1}^o \pm \frac{R_{k1}}{jf} & L_{k2}^o \pm \frac{R_{k2}}{jf} & \cdots & -\sum_{j=1}^{n} M_{kj} I_{cj} & \cdots & L_{km}^o \pm \frac{R_{km}}{jf} \\
\vdots & \vdots & \ddots & \vdots & \ddots & \vdots \\
L_{m1}^o \pm \frac{R_{m1}}{jf} & L_{m2}^o \pm \frac{R_{m2}}{jf} & \cdots & -\sum_{j=1}^{n} M_{mj} I_{cj} & \cdots & L_{mm}^o + \frac{R_m}{jf}
\end{vmatrix}.
\tag{3.30}
$$

Rewriting determinant (3.30) as follows:

$$
\Delta_k =
\begin{vmatrix}
L_{11}^o + \frac{R_1}{jf} & L_{21}^o \pm \frac{R_{21}}{jf} & \cdots & L_{k1}^o \pm \frac{R_{k1}}{jf} & \cdots & L_{m1}^o \pm \frac{R_{m1}}{jf} \\
L_{12}^o \pm \frac{R_{12}}{jf} & L_{22}^o + \frac{R_2}{jf} & \cdots & L_{k2}^o \pm \frac{R_{k2}}{jf} & \cdots & L_{m2}^o \pm \frac{R_{m2}}{jf} \\
\vdots & \vdots & \ddots & \vdots & \ddots & \vdots \\
-\sum_{j=1}^{n} M_{1j} I_{cj} & -\sum_{j=1}^{n} M_{2j} I_{cj} & \cdots & -\sum_{j=1}^{n} M_{kj} I_{cj} & \cdots & -\sum_{j=1}^{n} M_{mj} I_{cj} \\
\vdots & \vdots & \ddots & \vdots & \ddots & \vdots \\
L_{1m}^o \pm \frac{R_{1m}}{jf} & L_{2m}^o \pm \frac{R_{2m}}{jf} & \cdots & L_{km}^o \pm \frac{R_{km}}{jf} & \cdots & L_{mm}^o + \frac{R_m}{jf}
\end{vmatrix},
\tag{3.31}
$$

and accounting for the determinant properties, (3.30) can be represented as the sum:

$$
\Delta_k = -\sum_{j=1}^{N} \Delta_{kj} I_{cj},
\tag{3.32}
$$

where

$$
\Delta_{kj} = -
\begin{vmatrix}
L_{11}^o + \frac{R_1}{jf} & L_{21}^o \pm \frac{R_{21}}{jf} & \cdots & L_{k1}^o \pm \frac{R_{k1}}{jf} & \cdots & L_{n1}^o \pm \frac{R_{n1}}{jf} \\
L_{12}^o \pm \frac{R_{12}}{jf} & L_{22}^o + \frac{R_2}{jf} & \cdots & L_{k2}^o \pm \frac{R_{k2}}{jf} & \cdots & L_{n2}^o \pm \frac{R_{n2}}{jf} \\
\vdots & \vdots & \ddots & \vdots & \ddots & \vdots \\
M_{1j} & M_{2j} & \cdots & M_{kj} & \cdots & M_{nj} \\
\vdots & \vdots & \ddots & \vdots & \ddots & \vdots \\
L_{1n}^o \pm \frac{R_{1n}}{jf} & L_{2n}^o \pm \frac{R_{2n}}{jf} & \cdots & L_{kn}^o \pm \frac{R_{kn}}{jf} & \cdots & L_{nn}^o + \frac{R_{nn}}{jf}
\end{vmatrix}.
\tag{3.33}
$$

Taking the later equation into account, the current corresponding to the k-th eddy current circuit can be directly written in terms of the coils currents. Hence, Eq. (3.28) becomes

$$
I_k = \frac{-\sum_{j=1}^{N} \Delta_{kj} I_{cj}}{\Delta}.
\tag{3.34}
$$

Thus, instead of $m + 6$ equations, set (3.15) can be reduced to six equations as follows:

$$\begin{cases} m\ddot{q}_l + \mu_l\dot{q}_l - \dfrac{\partial W_m}{\partial q_l} = F_l; \ l = 1, 2; \\[3mm] m\ddot{q}_3 + \mu_3\dot{q}_3 + Mg - \dfrac{\partial W_m}{\partial q_3} = F_3; \\[3mm] J_l\ddot{q}_l + \mu_l\dot{q}_l - \dfrac{\partial W_m}{\partial q_l} = T_l; \ l = 4, 5, 6, \\[3mm] , \end{cases} \qquad (3.35)$$

where the derivative of W_m with respect to a generalized coordinate has the following general form:

$$\frac{\partial W_m}{\partial q_r} = \frac{1}{2}\sum_{k=1}^{n}\sum_{s=1}^{n}\frac{\partial L_{ks}^o}{\partial q_r}i_k i_s + \sum_{k=1}^{n}\sum_{j=1}^{N}\frac{\partial M_{kj}}{\partial q_r}i_k i_{cj}. \qquad (3.36)$$

Hence, the behaviour of the system is defined only by the generalized coordinates of its mechanical part. Moreover, the number of generalized coordinates of the mechanical part can be further reduced, depending on a particular design, as will be shown below.

Accounting for Eq. (3.34), the derivative of W_m with respect to a generalized coordinate (3.36) can be written via current amplitudes of coils as

$$\frac{\partial W_m}{\partial q_r} = \overbrace{\frac{1}{\Delta^2}\frac{1}{2}\sum_{k=1}^{n}\sum_{s=1}^{n}\frac{\partial L_{ks}^o}{\partial q_r}\sum_{j=1}^{N}\sum_{i=1}^{N}\Delta_{kj}\Delta_{si}I_{cj}I_{ci}}^{\text{I term}} - \underbrace{\frac{1}{\Delta}\sum_{k=1}^{n}\sum_{j=1}^{N}\frac{\partial M_{kj}}{\partial q_r}I_{cj}\sum_{s=1}^{N}\Delta_{ks}I_{cs}}_{\text{II term}}.$$

$$(3.37)$$

The analysis of Eq. (3.37) shows that two sources of ponderomotive forces can be identified: the one source due to changing in the shape of eddy current circuits and the positions of the these circuits with respect to each other within the micro-object (the first term in Eq. (3.37)) and another one due to changing the positions of the induced eddy current circuits with respect to the coils (the second term in Eq. (3.37)). The amplitudes of the eddy currents are several orders of magnitude less than the amplitudes of the coil currents. As a result, the ponderomotive force, which is defined by the first term in Eq. (3.37), is negligible compared to the second one. Hence, Eq. (3.37) can be simplified by keeping only the second term in the equation as follows:

$$\frac{\partial W_m}{\partial q_r} = -\underbrace{\frac{1}{\Delta}\sum_{k=1}^{n}\sum_{j=1}^{N}\frac{\partial M_{kj}}{\partial q_r}I_{cj}\sum_{s=1}^{N}\Delta_{ks}I_{cs}}_{\text{II term}}; \ r = 1,\ldots,6. \qquad (3.38)$$

For further analysis, the derivative of stored magnetic energy, W_m, with respect to the generalized coordinates, q_r, is expanded into the Taylor series. Due to the above-mentioned assumption of small displacements of the micro-object relative to the equilibrium position, the following functions taken from (3.38) can be expanded into the Taylor series as follows:

$$M_{kj} = m_0^{kj} + \sum_{l=1}^{6} m_l^{kj} q_l + \frac{1}{2} \sum_{r=1}^{6} \sum_{l=1}^{6} m_{rl}^{kj} q_r q_l, \tag{3.39}$$

where m_0^{kj}, m_l^{kj} and m_{rl}^{kj} are the constants defined at the equilibrium point. Determinants Δ and Δ_{ks} are complex values due to their definitions (3.29) and (3.33), respectively, and assumed to be expressed in terms of the inductances L_{ks}^o, M_{kj} and resistances R_k and R_{ks}.

Taking into account that

$$\frac{\partial M_{kj}}{\partial q_r} = m_r^{kj} + \sum_{l=1}^{6} m_{rl}^{kj} q_l, \tag{3.40}$$

Equation (3.38) can be linearized as follows:

$$\frac{\partial W_m}{\partial q_r} = \underbrace{-\frac{1}{\Delta} \sum_{k=1}^{n} \sum_{j=1}^{N} m_r^{kj} I_{cj} \sum_{s=1}^{N} \Delta_{ks} I_{cs}}_{\bar{c}_{r0}} + \sum_{l=1}^{6} \underbrace{\left(-\frac{1}{\Delta} \sum_{k=1}^{n} \sum_{j=1}^{N} m_{rl}^{kj} I_{cj} \sum_{s=1}^{N} \Delta_{ks} I_{cs} \right)}_{\bar{c}_{rl}} \cdot q_l,$$

$$\tag{3.41}$$

where the overbar denotes a complex quantity, $\bar{c}_{r0}$ is the force, while $\bar{c}_{rl}$ is the stiffness. Moreover, the following equality:

$$\bar{c}_{rl} = \bar{c}_{lr} \tag{3.42}$$

is held because of $m_{rl}^{kj} = m_{lr}^{kj}$.

Accounting for (3.41), set (3.15) can be rewritten as

$$\begin{cases} m\ddot{q}_l + \mu_l \dot{q}_l + \bar{c}_{l0} + \sum_{r}^{6} \bar{c}_{lr} q_r = F_l; \quad l = 1, 2; \\[2mm] m\ddot{q}_3 + \mu_3 \dot{q}_3 + Mg + \bar{c}_{30} + \sum_{r}^{6} \bar{c}_{3r} q_r = F_3; \\[2mm] J_l \ddot{q}_l + \mu_l \dot{q}_l + \bar{c}_{l0} + \sum_{r}^{6} \bar{c}_{lr} q_r = T_l; \quad l = 4, 5, 6, \end{cases} \tag{3.43}$$

where $\bar{c}_{l0}$ and $\bar{c}_{lr}$ ($l, r = 1, \ldots, 6$) are complex numbers, defined by (3.41). At the equilibrium point, the following coefficients must be equal to

$$\bar{c}_{30} = -mg; \quad \bar{c}_{l0} = 0; \quad l = 1, 2, 4, 5, 6. \tag{3.44}$$

Hence, the final linearized model describing dynamics of micromachined inductive levitation system becomes

$$\begin{cases} m\ddot{q}_l + \mu_l \dot{q}_l + \sum_{r}^{6} \bar{c}_{lr} q_r = F_l; \quad l = 1, 2, 3; \\ J_l \ddot{q}_l + \mu_l \dot{q}_l + \sum_{r}^{6} \bar{c}_{lr} q_r = T_l; \quad l = 4, 5, 6. \end{cases} \tag{3.45}$$

Generalized linear model (3.45) developed here, assuming small displacements of the levitated micro-object and its quasi-static behaviour, can now be applied to study the dynamics and stability of a micromachined inductive levitation system.

3.3 Stability of Inductive Levitation Systems

Let us represent linear model (3.45) in a matrix form as

$$\underline{A}\ddot{\bar{q}} + \underline{B}\dot{\bar{q}} + (\underline{R} + {}_J\underline{P})\bar{q} = f, \tag{3.46}$$

where $\bar{q} = [\bar{q}_1 \ \ldots \ \bar{q}_6]^T$ is the column-matrix of generalized coordinates, which are complex variables due to (3.45); $f = [F_1 \ F_2 \ F_3 \ T_4 \ T_5 \ T_6]^T$ is the column-matrix of generalized forces and torques applied to the micro-object; $\underline{A}=\text{diag}[m \ m \ m \ J_4 \ J_5 \ J_6]$ is the diagonal matrix of the micro-object mass and its moments of inertia; $\underline{B} = \text{diag}[\mu_1 \ \ldots \ \mu_6]$ is the diagonal matrix of damping coefficients; $\underline{R} = [\text{Re}\{\bar{c}_{lr}\}]$ and $\underline{P} = [\text{Im}\{\bar{c}_{lr}\}]$.

The physical meanings of matrices $\underline{A}$, $\underline{B}$ and $\underline{R}$ are obvious. Matrix $\underline{P}$ presents the coefficients of *circulatory forces* (the nonconservative positional forces) due to the dissipation of eddy current. Equation (3.46) can be rewritten by using only real values, and at the equilibrium point the linear model is equivalent to

$$\begin{bmatrix} A & 0 \\ 0 & A \end{bmatrix} \begin{bmatrix} \ddot{q} \\ \ddot{q}* \end{bmatrix} + \begin{bmatrix} B & 0 \\ 0 & B \end{bmatrix} \begin{bmatrix} \dot{q} \\ \dot{q}* \end{bmatrix} + \begin{bmatrix} R & -P \\ P & R \end{bmatrix} \begin{bmatrix} q \\ q* \end{bmatrix} = 0, \tag{3.47}$$

where $[q|q*]^T$ is the block column-vector of 12 variables; $q = \text{Re}\{\bar{q}\}$, $q* = \text{Im}\{\bar{q}\}$, $\underline{0}$ is a zero matrix and all block matrices have 12×12 elements. It is visible that

$$\left[\begin{array}{c|c} R & -P \\ \hline P & R \end{array}\right] = \left[\begin{array}{c|c} R & 0 \\ \hline 0 & R \end{array}\right] + \left[\begin{array}{c|c} 0 & -P \\ \hline P & 0 \end{array}\right], \tag{3.48}$$

and

$$\left[\begin{array}{c|c} 0 & -P \\ \hline P & 0 \end{array}\right] = -\left[\begin{array}{c|c} 0 & -P \\ \hline P & 0 \end{array}\right]^{T} \tag{3.49}$$

is a skew-symmetric matrix which corresponds to the positional nonconservative forces.

Analysis of model (3.46) reveals the following general issues related to stability of inductive levitation micro-systems, which are in particular formulated in terms of three theorems.

Theorem 1 (Unstable levitation I) *If a inductive levitation micro-system is subjected to only electromagnetic forces defined by (3.38) (without dissipation forces, so that $\underline{B} = 0$), then stable levitation in the micro-system is impossible.*

Proof According to the statement of the theorem, model (3.46) is rewritten as

$$\underline{A}\ddot{\bar{q}} + \left(\underline{R} + J\underline{P}\right)\bar{q} = 0. \tag{3.50}$$

Let us consider two cases. The first case is when matrix $\underline{R}$ is negative definite ($\underline{R} < 0$), and the second case is when matrix $\underline{R}$ is positive definite ($\underline{R} > 0$).

The case of $\underline{R} < 0$ is the trivial one, since system (3.50) becomes unstable. Due to the fact, adding the positional forces to such a system cannot support stable levitation [15, p. 203, Theorem 6.13]. For the case of $\underline{R} > 0$, system (3.50) can be transformed by introducing a new complex vector $\bar{u}$ such that

$$\bar{q} = \Lambda\bar{u}, \tag{3.51}$$

where Λ is the orthogonal matrix of transformation, matrices $\underline{A}$ and $\underline{R}$ can be represented as

$$\Lambda^{T}\underline{A}\Lambda = \underline{E}, \ \Lambda^{T}\underline{R}\Lambda = \underline{R}_{0}, \tag{3.52}$$

where $\underline{R}_{0} = \text{diag}(r_{1}, \ldots, r_{6})$ and $\underline{E}$ is the identity matrix. Accounting for (3.51) and (3.52), model (3.50) becomes

$$\underline{E}\ddot{\bar{u}} + \left(\underline{R}_{0} + J\hat{\underline{P}}\right)\bar{u} = 0, \tag{3.53}$$

where $\hat{\underline{P}} = \Lambda^{T}\underline{P}\Lambda$. Equation (3.53) can be rewritten in real values as

$$\left(\begin{array}{c|c} \underline{E} & 0 \\ \hline 0 & \underline{E} \end{array}\right)\left(\begin{array}{c} \ddot{u} \\ \ddot{u}* \end{array}\right) + \left(\begin{array}{c|c} \underline{R}_{0} & -\hat{\underline{P}} \\ \hline \hat{\underline{P}} & \underline{R}_{0} \end{array}\right)\left(\begin{array}{c} u \\ u* \end{array}\right) = 0, \tag{3.54}$$

for which the characteristic equation is

$$\det \left(\frac{\underline{I}\lambda^2 + \underline{R}_0}{\hat{\underline{P}}} \middle| \frac{-\hat{\underline{P}}}{\underline{I}\lambda^2 + \underline{R}_0} \right) = 0, \tag{3.55}$$

or

$$\det \left(\left(\underline{I}\lambda^2 + \underline{R}_0 \right)^2 + \hat{\underline{P}}^2 \right) = 0. \tag{3.56}$$

Due to the condition (3.42), the matrix $\underline{P}$ is symmetric. As a result the matrix $\hat{\underline{P}}^2 > 0$ becomes the positive definite matrix [16, p. 34], [17]. Hence, the following characteristic equation $\det \left(\tilde{\lambda}^2 + \hat{\underline{P}}^2 \right) = 0$ has 12 imaginary roots, $\tilde{\lambda}_i = \pm J a_i$, $a_i > 0$, $i = (1, \ldots, 6)$. Accounting for $\tilde{\lambda}_i = \lambda^2 + r_i$, where $r_i > 0$, $i = (1, \ldots, 6)$, the roots of (3.56) become

$$\begin{aligned} \lambda_i &= \pm J \sqrt{(r_i - J a_i)}, \quad i = (1, \ldots, 6), \\ \lambda_j &= \pm J \sqrt{(r_j + J a_j)}, \quad j = (7, \ldots, 12). \end{aligned} \tag{3.57}$$

Finally, we have

$$\begin{aligned} \lambda_i &= \pm \sqrt{\frac{\sqrt{r_i^2 + a_i^2} - r_i}{2}} \pm J \sqrt{\frac{\sqrt{r_i^2 + a_i^2} + r_i}{2}}, i = (1, \ldots, 6) \\ \lambda_j &= \mp \sqrt{\frac{\sqrt{r_j^2 + a_j^2} - r_j}{2}} \pm J \sqrt{\frac{\sqrt{r_j^2 + a_j^2} + r_j}{2}}, j = (7, \ldots, 12) \end{aligned} \tag{3.58}$$

From Eq. (3.58), it is seen that the real part of the roots have positive values. This fact proves the theorem.

This theorem can be referred to the main feature of inductive levitation systems. However, if the levitating micro-object is a perfect conductor, then $\underline{P} = 0$. Hence, when matrix $\underline{R}$ is the positive definite matrix, stable levitation without dissipative forces becomes possible. Also another conclusion can be formulated in the following corollary.

Corollary 1 *If a micromachined inductive levitation system is subjected to only electromagnetic forces, and the potential part of the electromagnetic forces is absent* $(\underline{R} = 0)$, *then stable levitation in the system is impossible.*

This fact follows directly from (3.58). Substituting $r_i = 0$ into (3.58), the roots still have a positive real part. Also, the corollary agrees with theorem [15, p. 197, Theorem 6.10] about the equilibrium of a system subjected only to positional forces.
Even if the dissipative forces are added to such a system without potential forces, the stable levitation is still impossible, this fact can be formulated in the second theorem below.

Theorem 2 (Unstable levitation II) *If a micromachined inductive levitation system is subjected to electromagnetic forces having only positional* $\underline{P} \neq 0$ $(\underline{R} = 0)$ *and dissipative forces* $(\underline{B} > 0)$, *then stable levitation is impossible.*

Proof We consider the following equation

$$\underline{A}\ddot{\bar{q}} + \underline{B}\dot{\bar{q}} + {}_J\underline{P}\bar{q} = 0. \tag{3.59}$$

As it was done above, the complex vector $\bar{u}$ given in (3.51) is used, hence, matrices $\underline{A}$ and $\underline{B}$ can be represented as follows:

$$\underline{\Lambda}^T \underline{A} \underline{\Lambda} = \underline{E}, \quad \underline{\Lambda}^T B \underline{\Lambda} = \underline{B}_0, \tag{3.60}$$

where $\underline{B}_0 = \mathrm{diag}(\hat{\mu}_1, \dots, \hat{\mu}_6)$. Taking later equations into account, Eq. (3.59) is rewritten as

$$\underline{E}\ddot{\bar{u}} + \underline{B}_0\dot{\bar{u}} + {}_J\hat{\underline{P}}\bar{u} = 0, \tag{3.61}$$

where $\hat{\underline{P}} = \underline{\Lambda}^T \underline{P}\underline{\Lambda}$. Equation (3.61) can be rewritten in real values as

$$\left(\begin{array}{c|c} E & 0 \\ \hline 0 & E \end{array}\right)\left(\begin{array}{c} \ddot{u} \\ \ddot{u}* \end{array}\right) + \left(\begin{array}{c|c} B_0 & 0 \\ \hline 0 & B_0 \end{array}\right)\left(\begin{array}{c} \dot{u} \\ \dot{u}* \end{array}\right) + \left(\begin{array}{c|c} 0 & -\hat{P} \\ \hline \hat{P} & 0 \end{array}\right)\left(\begin{array}{c} u \\ u* \end{array}\right) = 0. \tag{3.62}$$

The characteristic equation is

$$\det\left(\begin{array}{c|c} I\lambda^2 + B_0\lambda & -\hat{P} \\ \hline \hat{P} & I\lambda^2 + B_0\lambda \end{array}\right) = 0, \tag{3.63}$$

or

$$\det\left(\left(I\lambda^2 + B_0\lambda\right)^2 + \hat{P}^2\right) = 0. \tag{3.64}$$

Using the same reasoning as for Theorem 1, the roots are

$$\begin{array}{l} \lambda_i = \dfrac{-\hat{\mu}_i + \sqrt{\hat{\mu}_i^2 \mp {}_J 4a_i}}{2}, \ i = (1, \dots, 6) \\[2mm] \lambda_j = \dfrac{-\hat{\mu}_i - \sqrt{\hat{\mu}_i^2 \mp {}_J 4a_i}}{2}, \ j = (7, \dots, 12) \end{array} \tag{3.65}$$

Here, we need to prove that the real part of $\mathrm{Re}(\lambda_i) > 0$ is positive. Accounting for

$$\sqrt{\hat{\mu}_i^2 \mp {}_J 4a_i} = \sqrt{\dfrac{\sqrt{\hat{\mu}_i^4 + 16a_i^2} + \hat{\mu}_i^2}{2}} \mp {}_J\sqrt{\dfrac{\sqrt{\hat{\mu}_i^4 + 16a_i^2} - \hat{\mu}_i^2}{2}}, \tag{3.66}$$

the real part of λ_i is

$$\mathrm{Re}(\lambda_i) = \dfrac{1}{2}\left(-\hat{\mu}_i + \sqrt{\dfrac{\sqrt{\hat{\mu}_i^4 + 16a_i^2} + \hat{\mu}_i^2}{2}}\right). \tag{3.67}$$

We can write

$$-\hat{\mu}_i + \sqrt{\frac{\sqrt{\hat{\mu}_i^4 + 16a_i^2} + \hat{\mu}_i^2}{2}} > 0. \tag{3.68}$$

Inequality (3.68) is rewritten as

$$\sqrt{\hat{\mu}_i^4 + 16a_i^2} > \hat{\mu}_i^2, \tag{3.69}$$

which yields

$$16a_i^2 > 0. \tag{3.70}$$

This fact shows that the real part of λ_i is positive. Hence, the theorem is proved.

It is important to note that Theorem 2 corresponds to Theorem [15, p. 198, Theorem 6.11], which claims that the equilibrium of a system subjected to arbitrary nonconservative positional forces and linear dissipative forces is always unstable. The stable inductive levitation can be only achieved by adding the dissipative force. Upon holding the following necessary and sufficient conditions given in the theorem below, the system can be asymptotically stable.

Theorem 3 (Asymptotically stable levitation) *By adding dissipative forces ($\underline{B} > 0$) to a micromachined inductive levitation system subjected to electromagnetic forces defined by (3.38) and having a positive definite matrix of potential forces ($\underline{R} > 0$), the system can be asymptotically stable.*

Proof In order to prove the theorem, Metelitsyn's inequality [18] is used [19, p. 32]. The necessary condition is that matrix $\underline{A}$, $\underline{B}$ and $\underline{R}$ should be positive definite. The condition follows from the statement of the theorem. According to [20, p. 1099], a sufficient practical condition for asymptotically stable levitation for the present case becomes as follows:

$$\mu_{min} > p_{max}\sqrt{a_{max}/r_{min}}, \tag{3.71}$$

where μ_{min}, and r_{min} are the respective minimum values of $\underline{B}$ and $\underline{R}$; p_{max} and a_{max} are the respective maximum values of $\underline{P}$ and $\underline{A}$. This fact proves that the real part of eigenvalues is negative when the inequality (3.71) holds true. Thus, the system is asymptotically stable.

Operating IL-micro-systems in air, inequality (3.71) automatically holds due to the fact that damping forces dominate in the micro-world. Note that inequality (3.71) should be separately verified upon using IL-micro-systems in a vacuum environment.

3.4   Modelling of IL-Micro-Systems Based on Symmetric Designs

In this section, a generalization of the analytical approach developed above to analyse the dynamics and stability of several symmetric and axially symmetric designs of IL-micro-systems is considered.

A variety of axially symmetric designs of inductive levitation systems based on planar and 3D micro-coils are shown in Fig. 3.2. In particular, the IL-micro-system design shown in Fig. 3.2a was utilized in the prototype reported in [21] and proposed for its potential application as a gyroscope. The designs shown in Fig. 3.2b, c were employed in micro-gyroscope prototypes reported in [3, 4, 22] in which the rotation of a disc-shaped rotor was demonstrated. The design based on 3D micro-coils shown in Fig. 3.2d was realized in the prototype reported in [23]. Figure 3.2e shows the possible design of an IL-micro-system based on spiral-shaped 3D micro-coils in order to levitate, for instance, a conducting micro-sphere.

Examples of symmetric designs are shown in Fig. 3.3. The design shown in Fig. 3.3a was utilized in a prototype of accelerator for sorting micro-objects [24]. Figure 3.3b presents the design based on 3D micro-coils, which can be employed as a linear-transporter of micro-objects. The prototype based on this design will be demonstrated below in Chap. 5 and its stability will be studied theoretically and experimentally.

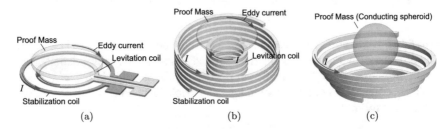

Fig. 3.2 Axially symmetric designs based on planar and 3D micro-coils: I is the electric current

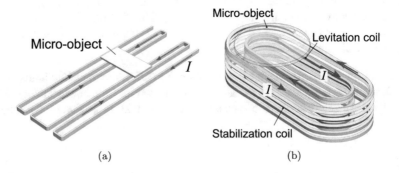

Fig. 3.3 Symmetric designs based on planar and 3D micro-coils: I is the electric current

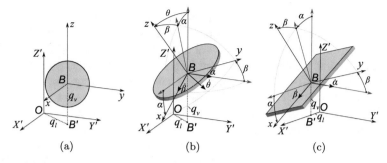

Fig. 3.4 Coordinate frames and generalized coordinates to define the position of spherical, disc and rectangle-shaped proof masses for axially symmetric and symmetric designs

Table 3.1 The structures of the analytical model as a function of design

Design	Levitating micro-object	Model structure
Axially symmetric		$\begin{cases} M\ddot{q}_v + \mu_v \dot{q}_v + \overline{c}_{vv} q_v = F_v; \\ M\ddot{q}_l + \mu_l \dot{q}_l + \overline{c}_{ll} q_l = F_l. \end{cases}$
		$\begin{cases} M\ddot{q}_v + \mu_v \dot{q}_v + \overline{c}_{vv} q_v = F_v; \\ M\ddot{q}_l + \mu_l \dot{q}_l + \overline{c}_{ll} q_l + \overline{c}_{l\theta}\theta = F_l; \\ J_\theta \ddot{\theta} + \mu_\theta \dot{\theta} + \overline{c}_{\theta l} q_l + \overline{c}_{\theta\theta}\theta = T_\theta. \end{cases}$
Symmetric		$\begin{cases} M\ddot{q}_v + \mu_v \dot{q}_v + \overline{c}_{vv} q_v = F_v; \\ M\ddot{q}_l + \mu_l \dot{q}_l + \overline{c}_{ll} q_l + \overline{c}_{l\alpha}\alpha = F_l; \\ J_\alpha \ddot{\alpha} + \mu_\alpha \dot{\alpha} + \overline{c}_{\alpha l} q_l + \overline{c}_{\alpha\alpha}\alpha = T_\alpha; \\ J_\beta \ddot{\beta} + \mu_\beta \dot{\beta} + \overline{c}_{\beta\beta}\beta = T_\beta. \end{cases}$

Due to the symmetry in the considered designs, the number of equations in set (3.45) can be reduced. For the case of axially symmetric designs and a spherical proof mass, the position of the levitated sphere is described by two generalized coordinates, namely, q_v and q_l representing the vertical and lateral linear displacements, as shown in Fig. 3.4a. Let us assign the origin of the coordinate frame $X'Y'Z'$ to the equilibrium point, O, in such a way that the Z' axis is parallel to the Z axis. The coordinate frame xyz is assigned to the mass centre of the proof mass. Then the generalized coordinate q_v characterizes the linear displacement of sphere's centre-of-mass, parallel to the Z' axis from the $X'Y'$ surface. The generalized coordinate q_l characterizes the linear displacement of the sphere's centre-of-mass on the $X'Y'$ surface from the O point. Hence, the model is reduced to a set of two equations. The behaviour of the disc-shaped proof mass without rotation can be described by three generalized coordinates [14]. In addition to the two linear coordinates, q_l and q_v, the angular generalized coordinate, θ is used as shown in Fig. 3.4b. For the symmetric designs shown in Fig. 3.3, it can be assumed that the levitated micro-object is in a neutral equilibrium state along the transportation line. Directing the Y' axis parallel to this line of transportation and locating the point, O, on the symmetry axis of the design, the generalized coordinates can be introduced as shown in Fig. 3.4c. The generalized coordinates q_l and q_v characterize the linear displacement of the micro-object along the X' axis and the vertical one parallel to the Z' axis, respectively, while two generalized coordinates α and β characterize its angular position.

Thus, depending on the design and the shape of a levitating micro-object, the model structures describing the behaviour of IL-micro-systems, in particular, the number of equations and elements of the complex matrix of the stiffness are already known from the defined generalized coordinates above and summarized in Table 3.1.

We suggest the following procedure for designing IL-micro-systems. Assuming that a micromachined inductive system is intended for using in air, the application of our approach is reduced to the analysis of the coefficients of matrix $\underline{R} = (\text{Re}\{\overline{c}_{lr}\})$, whose elements are defined in (3.41) as functions of the design parameters. A result of this analysis would be to find the domains of these design parameters where the matrix $\underline{R} > 0$ is positive definite, or to demonstrate that such domains do not exist ($\underline{R} < 0$ is everywhere negative definite). Additionally, for a vacuum environment, it becomes necessary to define the coefficients of the matrix $\underline{P} = (\text{Im}\{\overline{c}_{lr}\})$, which give the required values of the damping coefficients, μ_r, in order to fulfil the condition for stable levitation as defined in Theorem 3.

References

1. I. Ciric, Electromagnetic levitation in axially symmetric systems. Rev. Roum. Sci. Tech. Ser. Electrotech. Energ **15**(1), 35–73 (1970)
2. Z. Lu, F. Jia, J. Korvink, U. Wallrabe, V. Badilita, Design optimization of an electromagnetic microlevitation System based on copper wirebonded coils, in *2012 Power MEMS* (Atlanta, GA, USA 2012) pp. 363–366
3. C.B. Williams, C. Shearwood, P.H. Mellor, Modeling and testing of a frictionless levitated micromotor. Sensor. Actuat. A-Phys. **61**, 469–473 (1997)
4. W. Zhang, W. Chen, X. Zhao, X. Wu, W. Liu, X. Huang, S. Shao, The study of an electromagnetic levitating micromotor for application in a rotating gyroscope. Sensor. Actuat. A-Phys. **132**(2), 651–657 (2006)
5. K. Liu, W. Zhang, W. Liu, W. Chen, K. Li, F. Cui, S. Li, An innovative micro-diamagnetic levitation system with coils applied in micro-gyroscope. Microsyst. Technol. **16**(3), 431 (2009). https://doi.org/10.1007/s00542-009-0935-x
6. E. Laithwaite, Electromagnetic levitation. Proc. Inst. Electr. Eng. **112**(12), 2361–2375 (1965)
7. S. Babic, F. Sirois, C. Akyel, C. Girardi, Mutual inductance calculation between circular filaments arbitrarily positioned in space: alternative to Grovers formula. IEEE Trans. Magn. **46**, 3591–3600 (2010)
8. K.V. Poletkin, J.G. Korvink, Efficient calculation of the mutual inductance of arbitrarily oriented circular filaments via a generalisation of the Kalantarov-Zeitlin method. J. Magn. Magn. Mater. **483**, 10–20 (2019). https://doi.org/10.1016/j.jmmm.2019.03.078
9. K. Poletkin, A. Chernomorsky, C. Shearwood, U. Wallrabe, A qualitative analysis of designs of micromachined electromagnetic inductive contactless suspension. Int. J. Mech. Sci. **82**, 110–121 (2014). https://doi.org/10.1016/j.ijmecsci.2014.03.013
10. K. Poletkin, Z. Lu, U. Wallrabe, J. Korvink, V. Badilita, Stable dynamics of micro-machined inductive contactless suspensions. Int. J. Mech. Sci. **131–132**, 753–766 (2017). https://doi.org/10.1016/j.ijmecsci.2017.08.016

11. Y.G. Martynenko, *Analytical Dynamics of Electromechanical Systems* (Izd. Mosk. Ehnerg. Inst, Moscow, 1985)

12. V.I. Arnol'd, *Mathematical Methods of Classical Mechanics*, vol. 60 (Springer Science & Business Media, Berlin, 2013)

13. J.C. Maxwell, *A Treatise on Electricity and Magnetism*, vol. 2, 3 edn. (Dover Publications Inc., 1954)

14. K. Poletkin, A.I. Chernomorsky, C. Shearwood, U. Wallrabe, An analytical model of micromachined electromagnetic inductive contactless suspension, in *the ASME 2013 International Mechanical Engineering Congress & Exposition* (ASME, San Diego, California, USA 2013) pp. V010T11A072–V010T11A072 https://doi.org/10.1115/IMECE2013-66010

15. D.R. Merkin, in *Introduction to the Theory of Stability*, vol. 24, ed. by F.F. Afagh, A.L. Smirnov (Springer Science & Business Media, Berlin, 2012)

16. B.P. Demidovich, *Lectures on the Mathematical Theory of Stability* (Nauka, Moscow, 1967)

17. O.N. Kirillov, *Nonconservative Stability Problems of Modern Physics*, vol. 14 (Walter de Gruyter, 2013)

18. I. Metelitsyn, On gyroscopic stabilization. Dokl. Akad. Nauk SSSR **86**, 31–34 (1952)

19. A.P. Seyranian, A.A. Mailybaev, *Multiparameter Stability Theory with Mechanical Applications*, vol. 13 (World Scientific, 2003)

20. A. Seyranian, W. Kliem, Metelitsyn's inequality and stability criteria in mechanical problems, in *Physics and Control, 2003. Proceedings. 2003 International Conference*, vol. 4 (2003), pp. 1096–1101

21. C. Williams, C. Shearwood, P. Mellor, A. Mattingley, M. Gibbs, R. Yates, Initial fabrication of a micro-induction gyroscope. Microelectron. Eng. **30**(1–4), 531–534 (1996)

22. C. Shearwood, K.Y. Ho, C.B. Williams, H. Gong, Development of a levitated micromotor for application as a gyroscope. Sensor. Actuat. A-Phys. **83**(1–3), 85–92 (2000)

23. V. Badilita, M. Pauls, K. Kratt, U. Wallrabe, Contactless magnetic micro-bearing based on 3D solenoidal micro-coils for advanced powerMEMS components. Proc. of PowerMEMS **2009**, 87–90 (2009)

24. I. Sari, M. Kraft, A MEMS Linear Accelerator for Levitated Micro-objects. Sensor. Actuat. A-Phys. **222**, 15–23 (2015)

Chapter 4
Quasi-finite Element Modelling

In this chapter, the quasi-finite element approach for modelling of electromagnetic
levitation micro-systems is developed [1]. The developed approach allows calculating
accurately and efficiently a distribution of induced eddy current within a levitated
micro-object by means of using finite elements. The combination of finite element
manner to calculate induced eddy current and the set of six differential equations
describing the behaviour of the mechanical part of electromagnetic levitation system
is the resulting essence of the proposed quasi-finite element approach to simulate IL-
micro-systems. This fact makes the main difference between the proposed quasi-finite
element modelling and analytical one discussed in the previous chapter, in which the
distribution of eddy current within the levitated micro-object was assumed to be
known so that the entire eddy current distribution can be represented by selectively
chosen m eddy current circuits.

4.1 Statement of Problem

Let us consider a general design configuration of an electromagnetic levitation system
as shown in Fig. 4.1, which consists of N-arbitrary shaped wire loops and a levitated
conducting object. Each j-th wire loop is fed by its own ac current denoted by i_j
with the index corresponding to the index of the wire loop. The set of wire loops
generates the alternating magnetic flux passing through the levitated object. In turn,
the eddy current is induced within the conducting object. Force interaction between
induced eddy current and currents in wire loops provides the compensation of the
gravity force acting on the conducting object along X_3-axis of an inertial frame $\{X_i\}$
$(i = 1, 2, 3)$ and levitates then it sably at an equilibrium position. Assuming that the
levitated object is a rigid body, then its equilibrium position can be defined through
the six generalized coordinates corresponding to the three translation coordinates

K. Poletkin, *Levitation Micro-Systems*, Microsystems and Nanosystems,
https://doi.org/10.1007/978-3-030-58908-0_4

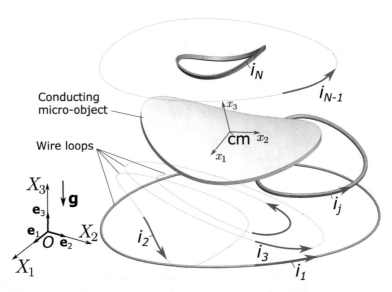

Fig. 4.1 The generalized scheme of a levitated micro-object by a system of arbitrary shaped wire loops with ac currents: $X_1 X_2 X_3$ is the fixed coordinate frame with the corresponding unit vectors $\mathbf{e}_1$, $\mathbf{e}_2$, and $\mathbf{e}_3$, respectively; $\mathbf{g}$ is the gravity acceleration vector directed along the X_3-axis; $x_1 x_2 x_3$ is the coordinate frame assigned to object principal axes with the origin at its centre of mass (cm); i_j is the current in the j-wire loop

and three angular ones with respect to the fixed frame $\{X_i\}$ $(i = 1, 2, 3)$ with the vector base $\boldsymbol{e}^X$.

In order to specify these six generalized coordinates, the coordinate frame $\{x_k\}$ $(k = 1, 2, 3)$ with the base $\underline{e}^x$ and corresponding unit vectors e_k^x $(k = 1, 2, 3)$ is rigidly attached to the levitated object, in such a way, that its origin is located at the centre of mass of the object as shown in Fig. 4.1. Also, axes of the coordinate frame $\{x_k\}$ $(k = 1, 2, 3)$ coincide with principal axes of inertia of the micro-object. The translational position of the micro-object centre mass (cm) with respect to the fixed frame is characterized by coordinates q_l $(l = 1, 2, 3)$ and the angular one by the Brayn angles (Cardan angles) denoted as q_l $(l = 4, 5, 6)$ similar to Sect. 3.2. Thus, six coordinates q_l $(l = 1 \ldots 6)$ can be considered as independent *generalized coordinates* of mechanical part of the electromagnetic levitation system.

A circuit of any shape can be replaced by a network of small current loops. For instance, we have a loop of current, I, as shown in Fig. 4.2. Imagining that this loop is filled by a surface. Then on the surface, a number of small loops is marked out, each of which can be considered as a plane. Now if the current, I, flows around each of small loops, the resulting current will be the same as the original loop current since the currents will cancel on the all lines internal to the loop. This point provides the reasoning for a further step.

Now let us assume that the condition of quasistationarity is held [2, p. 7], [3, p. 493]. Since the system of small currents is equivalent to the current circuit, induced

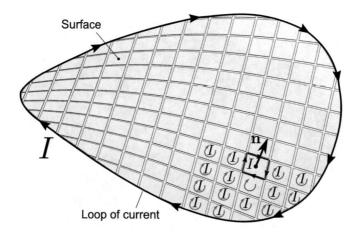

Fig. 4.2 The loop of current, I, can be considered as a network of small current loop: n is a unit normal to a surface

eddy current within the micro-object can be represented by n-electric circuits (finite elements), each of them consists of the inductor and resistor connecting in series [4]. Meshing the levitated micro-object by n-elements having the circle shape of the same radius as shown in Fig. 4.3, we denote i_k ($k = 1, 2, \ldots, n$) as the induced eddy current corresponding to the k-element, which can be represented as the *generalized velocities* of electromagnetic part of the levitation system. The linear position of s-circular element with respect to the coordinate frame $\{x_k\}$ ($k = 1, 2, 3$) is defined by the vector $^{(s)}\rho$, but the angular position of the same element is determined by Brayn angles, namely, $^{(s)}\phi = [^{(s)}\phi_1 \; ^{(s)}\phi_2 \; 0]^T$.

The circular shape of finite element is chosen due to the fact that it simplifies significantly the further calculation. Although such a shape of element introduces an error in the estimation of the value of eddy current magnitude, however this error in the calculation is eliminated by an equilibrium condition.

Adapting the generalized coordinates and the assumptions introduced above, the model can be written by using the Lagrange- Maxwell equations in the same was as set (3.15).

Similar to (4.1), the kinetic energy is

$$K = \frac{1}{2} \sum_{l=1}^{3} m\dot{q}_l^2 + \frac{1}{2} \sum_{l=1}^{3} J_l\dot{\omega}_l^2. \tag{4.1}$$

According to Fig. 4.1, the potential energy can be defined simply as follows:

$$\Pi = mgq_3. \tag{4.2}$$

The dissipation function is

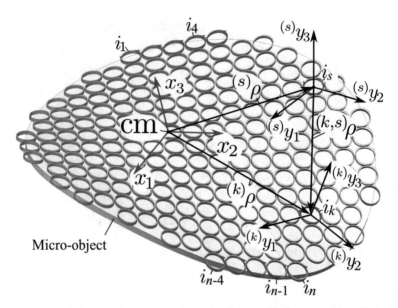

Fig. 4.3 The conducting micro-object is meshed by finite elements of circular shape: $\{^{(s)}y_l\}$ and $\{^{(k)}y_l\}$ ($l = 1, 2, 3$) are the coordinate frames assigned to s- and k-circular element, respectively

$$\Psi = \frac{1}{2}\sum_{l=1}^{3}\mu_l \dot{q}_l^2 + \frac{1}{2}\sum_{l=1}^{3}\nu_l \dot{\varphi}_l^2 + \frac{1}{2}R\sum_{k=1}^{n}i_k^2, \tag{4.3}$$

where μ_l and ν_l ($l = 1, 2, 3$) are the damping coefficients; R is the electrical resistance of the element. The energy stored within the electromagnetic field can be written as

$$
\begin{aligned}
W_m = {} & \frac{1}{2}\sum_{j=1}^{N}\sum_{s=1}^{N}L_{js}^c i_{cj}i_{cs} + \frac{1}{2}\sum_{k=1}^{n}\sum_{s=1}^{n}L_{ks}^o i_k i_s \\
& + \sum_{k=1}^{n}\sum_{j=1}^{N}M_{kj}i_k i_{cj},
\end{aligned}
\tag{4.4}
$$

where L_{jj}^c is the self-inductance of the j-wire loop; L_{js}^c, $j \neq s$ is the mutual inductance between j- and s-coils; $L_{kk}^o = L^o$ is the self-inductance of the circular element; $L_{ks}^o = L_{ks}^o(^{(k,s)}\rho, {}^{(k,s)}\phi)$, $k \neq s$ is the mutual inductance between k- and s-finite circular elements (besides, $^{(k,s)}\rho = {}^{(k)}\rho - {}^{(s)}\rho$ and $^{(k,s)}\phi = {}^{(k)}\phi - {}^{(s)}\phi$, see Fig. 4.3); $M_{kj} = M_{kj}(\underline{q})$ is the mutual inductance between the k-circular element and the j-wire loop.

As it was shown in Chap. 3 (also see works [4, 5]) upon assuming quasi-static behaviour of the micro-object, the induced eddy currents i_k ($k = 1, \ldots, n$) in circular elements can be directly expressed in terms of coil currents i_{cj} ($j = 1, \ldots, N$). Also, assuming that for each coil, the current i_{cj} is a periodic signal with an amplitude of I_{cj} (in general the amplitude is assumed to be a complex value) at the same frequency ω, we can write $i_{cj} = I_{cj}e^{J\omega t}$. Then, the $k-$ eddy current can be represented as $i_k = I_k e^{J\omega t}$, where I_k is the amplitude. Hence, according to [4, 5] first n equations of set (3.15) can be solved and the induced eddy current per circular element becomes a solution of the following set of linear equations:

$$\left[L^o + R/(Jf)\right]I_k + \sum_{s=1,\,s\neq k}^{n} L^o_{ks}I_s = -\sum_{j=1}^{N} M_{kj}I_{cj};$$
$$k = 1, \ldots, n. \tag{4.5}$$

Combining the set (4.5) with the last six equations of (3.15) the quasi-finite element model of inductive levitation system is obtained. Thus, the obtained quasi-FEM is the combination of finite element manner to calculate induced eddy current and the set of six differential equations describing the behaviour of mechanical part of electromagnetic levitation system.

4.2 Procedure for the Analysis of IL-Micro-Systems

Based on the proposed quasi-FEM the following procedure for the analysis and study of induction levitation micro-systems can be suggested. At the beginning, the levitated micro-object is meshed by circular elements of the same radius, R_e, as shown in Fig. 4.3, a value of which is defined by a number of elements, n. The result of meshing becomes a list of elements $\{^{(s)}\underline{C} = [^{(s)}\rho\ ^{(s)}\phi]^T\}$ $(s = 1, \ldots, n)$ containing information about a radius vector and an angular orientation for each element with respect to the coordinate frame $\{x_k\}$ $(k = 1, 2, 3)$. Now a matrix corresponding to the left side of equation (4.5) can be formed as follows:

$$\underline{L} = \left[L^o + R/(Jf)\right]\underline{E} + \underline{M}^o, \tag{4.6}$$

where $\underline{E}$ is the $(n \times n)$ unit matrix, $\underline{M}^o$ is the $(n \times n)$-symmetric hollow matrix whose elements are L^o_{ks} $(k \neq s)$. The self-inductance of the circular element is calculated by the known formula for a circular ring of circular cross-section

$$L^o = \mu_0 R_e \left[\ln 8/\varepsilon - 7/4 + \varepsilon^2/8 (\ln 8/\varepsilon + 1/3)\right], \tag{4.7}$$

where μ_0 is the magnetic permeability of free space, $\varepsilon = th/(2R_e)$, th is the thickness of a mashed layer of micro-object. It is recommended that the parameter ε is selected to be not larger 0.1. Elements of the $\underline{M}^o$ matrix, L^o_{ks} $(k \neq s)$ can be calculated by the formulas developed in Sect. 4.3.2 by means of the Kalantarov-Zeitlin method (also see works [6, 7]). Using the list of elements $\{^{(s)}\underline{C}\}$, we can estimate L^o_{ks} as follows $L^o_{ks} = L^o_{ks}\left(^{(k,s)}\underline{C}\right)$, where $^{(k,s)}\underline{C} = ^{(k)}\underline{C} - ^{(s)}\underline{C}$.

Then, we assign to each coil the coordinate frame $\{^{(j)}z_k\}$ ($k = 1, 2, 3$) with the corresponding base $^{(j)}\underline{e}^z$. The linear and angular position of $\{^{(j)}z_k\}$ ($k = 1, 2, 3$) with respect to the fixed frame $\{X_k\}$ ($k = 1, 2, 3$) is defined by the radius vector $^{(j)}r_c^X$ and the Bryan angles $^{(j)}\phi_c$ ($j = 1, \ldots, N$), respectively. Knowing a projection of the j-coil filament loop on the $^{(k)}y_1$-$^{(k)}y_2$ plane of the k-circular element, again the Kalantarov-Zeitlin method [7] can be used in order to calculate the mutual inductance. Hence, the ($n \times N$) matrix $\underline{M}_c$ consisting of elements of the mutual inductance M_{kj} can be obtained. The induced eddy current in each circular element is a solution of the following matrix equation:

$$I = \underline{L}^{-1}\underline{M}_c\underline{I}_c,\tag{4.8}$$

where $\underline{I}$ is the ($n \times 1$) matrix of eddy currents and $\underline{I}_c = [I_{c1} I_{c2} \ldots I_{cN}]^T$ is the given ($N \times 1$) matrix of currents in coils.

Finally, substituting the eddy current $\underline{I}$ into the stored electromagnetic energy (4.4) the ponderomotive forces acting on the levitated object can be found as the first derivative of the stored electromagnetic energy with respect to the generalized coordinates of mechanical part. Thus, in a matrix form we can write as follows:

$$\begin{aligned}
F_l &= \frac{\partial W_m}{\partial q_l} = \underline{I}^T \frac{\partial \underline{M}_c}{\partial q_l} \underline{I}_c; \; l = 1, 2, 3; \\
T_l &= \frac{\partial W_m}{\partial q_l} = \underline{I}^T \frac{\partial \underline{M}_c}{\partial q_l} \underline{I}_c; \; l = 4, 5, 6.
\end{aligned}\tag{4.9}$$

Below in Chap. 6, the quasi-FEM is applied to study linear and angular pull-in actuation in hybrid levitation micro-system.

4.3   Calculation of the Mutual Inductance of Circular Filaments

Here in this section, the Kalantarov-Zeitlin method for calculation of the mutual inductance between two circular filaments arbitrarily positioning with respect to each other is introduced. Let us consider two circular filaments having radii of R_p and R_s for the primary circular filament (the primary circle) and the secondary circular filament (the secondary circle), respectively, to be arbitrarily positioned in space, namely, they have a linear and angular misalignment, as is shown in Fig. 4.4. Also, let us assign a coordinate frame (CF) denoted as XYZ to the primary circle in such way that the Z axis is coincident with the circle axis and the XOY plane of the CF lies on the circle's plane, where the origin O corresponds to the centre of primary circle. In turn, the xyz CF is assigned to the secondary circle in a similar way so that its origin B is coincident with the centre of the secondary circle.

The linear position of the secondary circle with respect to the primary one is defined by the coordinates of the centre B (x_B, y_B, z_B). The angular position of

Fig. 4.4 General scheme of arbitrarily positioning two circular filaments with respect to each other: P is an arbitrary point on the secondary filament

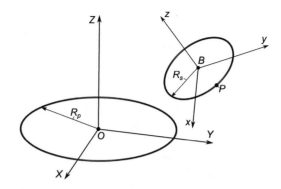

the secondary circle can be defined in two ways. Firstly, the angular position is defined by the angle θ and η corresponding to the angular rotation around an axis passing through the diameter of the secondary circle, and then the rotation of this axis lying on the surface $x'By'$ around the vertical z' axis, respectively, as it is shown in Fig. 4.5a. These angles for the determination of angular position of the secondary circle was proposed by Grover and used in his formula numbered by (179) in [8, p. 207] addressing the general case for calculation of the mutual inductance between two circular filaments.

The same angular position can be determined through the α and β angle, which corresponds to the angular rotation around the x' axis and then around the y'' axis, respectively, as it is shown in Fig. 4.5b. This additional second manner is more convenient in a case of study dynamics and stability, for instance, applying to axially symmetric inductive levitation systems [4, 9] in compared with the Grover manner. These two pairs of angles have the following relationship with respect to each other such as follows:

$$\begin{cases} \sin \beta = \sin \eta \sin \theta; \\ \cos \beta \sin \alpha = \cos \eta \sin \theta. \end{cases} \qquad (4.10)$$

The details of the derivation of this set presented above are shown in Appendix B.2.

4.3.1 The Kalantarov-Zeitlin Method

Using the general scheme for two circular filaments shown in Fig. 4.6 as an illustrative one, the Kalantarov-Zeitlin method is presented. The method reduces the calculation of mutual inductance between a circular primary filament and any other secondary filament having an arbitrary shape and any desired position with respect to the primary circular filament to a line integral [10, Sect. 1–12, p. 49].

Indeed, let us choose an arbitrary point P of the secondary filament (as it has been mentioned above the filament can have any shape), as shown in Fig. 4.6. An element of length $d\ell''$ of the secondary filament at the point P is considered. Also, the point

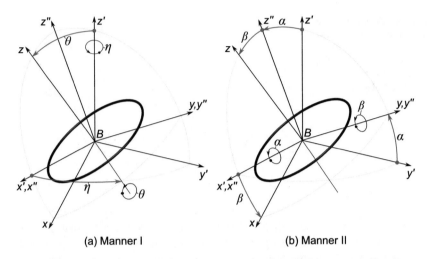

(a) Manner I   (b) Manner II

Fig. 4.5 Two manners for determining the angular position of the secondary circle with respect to the primary one: $x'y'z'$ is the auxiliary CF the axes of which are parallel to the axes of XYZ, respectively; $x''y''z''$ is the auxiliary CF defined in such a way that the x' and x'' are coincided, but the z'' and y'' axis is rotated by the α angle with respect to the z' and y' axis, respectively

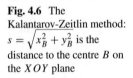

Fig. 4.6 The Kalantarov-Zeitlin method: $s = \sqrt{x_B^2 + y_B^2}$ is the distance to the centre B on the XOY plane

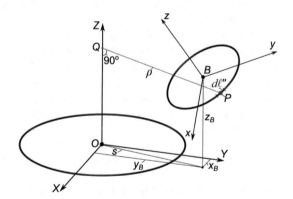

P is connected the point Q lying on the Z axis by a line, which is perpendicular to the Z axis and has a length of ρ, as shown in Fig. 4.6. Then the element $d\ell''$ can be decomposed on dz along the Z axis and on $d\rho$ along the ρ line and $d\lambda$ along the λ-circle having radius of ρ (see, Fig. 4.7). It is obvious that the mutual inductance between dz and the primary circular filament is equal to zero because dz is perpendicular to a plane of primary circle. But the mutual inductance between $d\rho$ and the primary circular filament is also equal to zero because of the symmetry of the primary circle relative to the $d\rho$ direction.

Thus, the mutual inductance dM between element $d\ell''$ and the primary circle is equal to the mutual inductance dM_λ between element $d\lambda$ and the primary circle. Moreover, due to the fact that the primary and the λ-circle are coaxial and, conse-

Fig. 4.7 The Kalantarov-Zeitlin method: projection of the secondary filament on the ρ-plane passed through the point P and parallel to the plane of the primary circular filament; $d\ell$ is the projection of the element $d\ell''$ on the ρ-plane

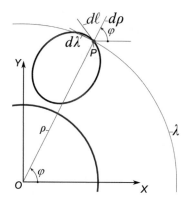

quently, symmetric, we can write:

$$\frac{dM_\lambda}{M_\lambda} = \frac{d\lambda}{\lambda} = \frac{d\lambda}{2\pi\rho}, \tag{4.11}$$

where M_λ is the mutual inductance of the primary and λ-circle.

From Fig. 4.7, it is directly seen that

$$d\lambda = dy\cos\varphi - dx\sin\varphi = (\cos\zeta\cos\varphi - \cos\varepsilon\sin\varphi)d\ell, \tag{4.12}$$

where $\cos\varepsilon$ and $\cos\zeta$ are the direction cosines of element $d\ell$ relative to the X and Y axis, respectively. Hence, accounting for (4.11) and (4.12), we can write:

$$dM = dM_\lambda = M_\lambda\frac{\cos\zeta\cos\varphi - \cos\varepsilon\sin\varphi}{2\pi\rho}d\ell, \tag{4.13}$$

and as a result, a line integral for calculation mutual inductance between the primary circle and a filament is

$$M = \frac{1}{2\pi}\int_\ell M_\lambda\frac{\cos\zeta\cos\varphi - \cos\varepsilon\sin\varphi}{\rho}d\ell, \tag{4.14}$$

where M_λ is defined by the Maxwell formula for mutual inductance between two coaxial circles [11, p. 340, Art. 701]. Note that during integrating, the Z coordinate of the element $d\ell$ is also changing and this dependency is taken into account by the M_λ function directly.

Fig. 4.8 Determination of the position of the point P on the ρ-plane through the fixed parameter s and the distance r

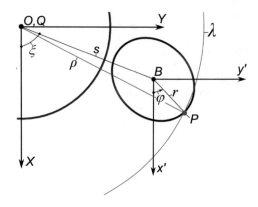

4.3.2 Derivation of Formulas

Due to the particular geometry of secondary filament under consideration, its projection on the ρ-plane (the ρ-plane is parallel to the primary circle plane and passed through the point P) is an ellipse, which can be defined in a polar coordinate by a function $r = r(\varphi)$ with the origin at the point B as it is shown in Fig. 4.8. Hence, the distance ρ can be expressed in terms of the parameter s, which is fixed, and the distance r from the origin B, which is varied with the angular variable φ. Introducing the angle γ as shown in Fig. 4.9, for the distance ρ the following equations can be written:

$$\rho \cos \gamma = r + s \cos(\xi - \varphi),$$
$$\rho \sin \gamma = s \sin(\xi - \varphi). \tag{4.15}$$

Due to (4.15), we have:

$$\rho^2 = r^2 + r \cdot s \cos(\xi - \varphi) + s^2, \tag{4.16}$$

where the function $r = r(\varphi)$ can be defined as

$$r = \frac{R_s \cos \theta}{\sqrt{\sin^2(\varphi - \eta) + \cos^2 \theta \cos^2(\varphi - \eta)}}. \tag{4.17}$$

The angle θ and η defines the angular position of the secondary circle with respect to the primary one according to manner I considered above. Note that the function r can be also defined via the angles α and β of manner II and as it is shown in Appendix . However, the further derivation, the angular position of the secondary circle is defined through manner I since it is convenient for the direct comparison with Grover's and Babič' results.

According to Fig. 4.9, the relationship between the element $d\lambda$ of the λ-circle and an increment of the angle φ is as follows:

Fig. 4.9 The relationship between $d\lambda$ and $d\varphi$

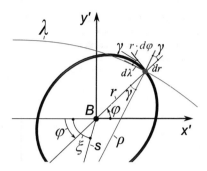

$$d\lambda = r \cdot d\varphi \cos\gamma - dr \sin\gamma = \left(r \cos\gamma - \frac{dr}{d\varphi} \sin\gamma\right) d\varphi. \qquad (4.18)$$

Then, accounting for (4.18), (4.16) and (4.15), line integral (6.4) can be replaced by a definite integral for the calculation of mutual inductance as follows:

$$M = \frac{1}{2\pi} \int_0^{2\pi} M_\lambda \frac{r^2 + r \cdot s \cos(\xi - \varphi) - \frac{dr}{d\varphi} s \sin(\xi - \varphi)}{\rho^2} d\varphi. \qquad (4.19)$$

Now, let us introduce the following dimensionless parameters such as follows:

$$\bar{x}_B = \frac{x_B}{R_s}; \ \bar{y}_B = \frac{y_B}{R_s}; \ \bar{z}_B = \frac{z_B}{R_s}; \ \bar{r} = \frac{r}{R_s};$$
$$\bar{\rho} = \frac{\rho}{R_s}; \ \bar{s} = \sqrt{\bar{x}_B^2 + \bar{y}_B^2}. \qquad (4.20)$$

The φ-derivative of $\bar{r}$ is

$$\frac{d\bar{r}}{d\varphi} = \frac{1}{2}\bar{r}^3 \tan^2\theta \sin(2(\varphi - \eta)), \qquad (4.21)$$

The mutual inductance M_λ [12, p. 6] is

$$M_\lambda = \mu_0 \frac{2}{k} \Psi(k) \sqrt{R_p R_s \bar{\rho}}, \qquad (4.22)$$

where μ_0 is the magnetic permeability of free space, and

$$\Psi(k) = \left(1 - \frac{k^2}{2}\right) K(k) - E(k), \qquad (4.23)$$

where $K(k)$ and $E(k)$ are the complete elliptic functions of the first and second kinds, respectively, and

Fig. 4.10 The special case: the two filament circles are mutually perpendicular to each other

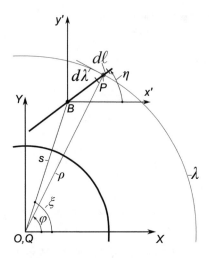

$$k^2 = \frac{4\nu\bar\rho}{(\nu\bar\rho + 1)^2 + \nu^2\bar z_\lambda^2}, \tag{4.24}$$

where $\nu = R_s/R_p$ and $\bar z_\lambda = \bar z_B + \bar r \tan\theta \sin(\varphi - \eta)$. Accounting for dimensionless parameters (4.20) and substituting (4.21) and (4.23) into integral (4.19), the new formula to calculate the mutual inductance between two circular filaments having any desired position with respect to each other becomes

$$M = \frac{\mu_0\sqrt{R_p R_s}}{\pi} \int_0^{2\pi} \frac{\bar r + t_1 \cdot \cos\varphi + t_2 \cdot \sin\varphi}{k\bar\rho^{1.5}} \cdot \bar r \cdot \Psi(k)d\varphi, \tag{4.25}$$

where terms t_1 and t_2 are defined as

$$\begin{aligned} t_1 &= \bar x_B + 0.5\bar r^2 \tan^2\theta \sin(2(\varphi - \eta)) \cdot \bar y_B; \\ t_2 &= \bar y_B - 0.5\bar r^2 \tan^2\theta \sin(2(\varphi - \eta)) \cdot \bar x_B, \end{aligned} \tag{4.26}$$

and $\bar\rho = \sqrt{\bar r^2 + 2\bar r \cdot \bar s \cos(\xi - \varphi) + \bar s^2}$.

Formula (4.25) can be applied to any possible cases, but one is excluded when the two filament circles are mutually perpendicular to each other. In this case the projection of the secondary circle onto the ρ-plane becomes simply a line as it is shown in Fig. 4.10 and, as a result, to integrate with respect to φ is no longer possible.

For the treatment of this case, the Kalantarov-Zeitlin formula (6.4) is directly used. Let us introduce the dimensionless variable $\bar\ell = \ell/R_s$ and then the integration of (6.4) is preformed with respect to this dimensionless variable $\bar\ell$ within interval $-1 \le \bar\ell \le 1$. The direction cosines $\cos\zeta$ and $\cos\varepsilon$ become as $\sin\eta$ and $\cos\eta$, respectively (see, Fig. 4.10). Accounting for

$$\rho \cos \varphi = s \cos \xi + \ell \cos \eta,$$
$$\rho \sin \varphi = s \sin \xi + \ell \sin \eta,$$

(4.27)

and the Maxwell formula (4.23) and (4.24), where the Z-coordinate of the element $d\bar{\ell}$ is defined as

$$\bar{z}_\lambda = \bar{z}_B \pm \sqrt{1 - \bar{\ell}^2},$$

(4.28)

then the formula to calculate the mutual inductance between two filament circles, which are mutually perpendicular to each other, becomes as follows:

$$M = \frac{\mu_0 \sqrt{R_p R_s}}{\pi} \left[\int_{-1}^{1} \frac{t_1 - t_2}{k \bar{\rho}^{1.5}} \cdot \Psi(k) d\bar{\ell} \right.$$
$$\left. + \int_{1}^{-1} \frac{t_1 - t_2}{k \bar{\rho}^{1.5}} \cdot \Psi(k) d\bar{\ell} \right],$$

(4.29)

where terms t_1 and t_2 are defined as

$$t_1 = \sin \eta (\bar{x}_B + \bar{\ell} \cos \eta);$$
$$t_2 = \cos \eta (\bar{y}_B + \bar{\ell} \sin \eta),$$

(4.30)

and $\bar{\rho} = \sqrt{s^2 + 2\bar{\ell} \cdot \bar{s} \cos(\xi - \eta) + \bar{\ell}^2}$. Note that integrating (4.29) between -1 and 1 Eq. (4.28) is calculated with the positive sign and for the other direction the negative sign is taken.

If parameter k trends to one. In this case formula (4.22) can be approximated by the following simple function [10, p. 466], which is

$$M_\lambda = \mu_0 r_c \left[\ln \frac{8r_c}{\sqrt{l_\lambda^2 + (\rho - r_c)^2}} - 2.0 \right].$$

(4.31)

In addition to, we can write the following relationship

$$\frac{r(r + s \cos \alpha)}{\rho^2} \approx \frac{1 + \frac{s}{r} \cos \alpha}{1 + 2\frac{s}{r} \cos \alpha} \approx 1.$$

(4.32)

Then, accounting for (4.31) and (4.32), the final simplified formula of the function of generalized coordinates for the mutual induction between the PM and coil becomes

$$M_m = \frac{\mu_0 r_c}{\pi} \int_0^\pi \left[\ln \frac{8r_c}{\sqrt{l_\lambda^2 + (\rho - r_c)^2}} - 2.0 \right] d\alpha,$$

(4.33)

where $\quad \rho = \rho(s) \approx \sqrt{r_{pm}^2 + 2r_{pm} \cdot s \cos \alpha + s^2} \quad$ and $\quad l_\lambda = l_\lambda(l, \varphi) = l - r_{pm}$ $\sin \varphi \cos \alpha$. The later formula is very useful and applicable for the analytical modell and analysis of stability and pull-in phenomenon in axially symmetric designs.

The *Matlab* codes of the implemented formulas (4.25) and (4.29) are available in Appendix . Also, in Appendix the developed formulas can be rewritten through the pair of the angle α and β.

References

1. K. Poletkin, Static pull-in behavior of hybrid levitation micro-actuators: simulation, modelling and experimental study. IEEE/ASME Trans. Mechatron. 1-1, (2020)
2. Y.G. Martynenko, *Analytical Dynamics of Electromechanical Systems* (Izd. Mosk. Ehnerg. Inst, Moscow, 1985)
3. B.G. Levich, *Theoretical Physics: An Advanced Text* (North-Holland, Amsterdam, 1971)
4. K. Poletkin, Z. Lu, U. Wallrabe, J. Korvink, V. Badilita, Stable dynamics of micro-machined inductive contactless suspensions. Int. J. Mech. Sci. **131–132**, 753–766 (2017). https://doi.org/10.1016/j.ijmecsci.2017.08.016
5. K. Poletkin, A.I. Chernomorsky, C. Shearwood, U. Wallrabe, An analytical model of micro-machined electromagnetic inductive contactless suspension, in *the ASME 2013 International Mechanical Engineering Congress & Exposition* (ASME, San Diego, California, USA 2013), pp. V010T11A072–V010T11A072. https://doi.org/10.1115/IMECE2013-66010
6. S. Babic, F. Sirois, C. Akyel, C. Girardi, Mutual inductance calculation between circular filaments arbitrarily positioned in space: alternative to Grovers formula. IEEE Trans. Magn. **46**, 3591–3600 (2010)
7. K.V. Poletkin, J.G. Korvink, Efficient calculation of the mutual inductance of arbitrarily oriented circular filaments via a generalisation of the Kalantarov-Zeitlin method. J. Magn. Magn. Mater. **483**, 10–20 (2019). https://doi.org/10.1016/j.jmmm.2019.03.078
8. F. Grover, *Inductance Calculations: Working Formulas and Tables* (Dover publications, Chicago, 2004)
9. K. Poletkin, A. Chernomorsky, C. Shearwood, U. Wallrabe, A qualitative analysis of designs of micromachined electromagnetic inductive contactless suspension. Int. J. Mech. Sci. **82**, 110–121 (2014). https://doi.org/10.1016/j.ijmecsci.2014.03.013
10. P.L. Kalantarov, L.A. Zeitlin, *Raschet induktivnostey (Calculation of Inductors)* (Energoatomizdat, Leningrad 1986), p. 488 (in Russian)
11. J.C. Maxwell, *A Treatise on Electricity and Magnetism*, vol. 2, 3 edn. (Dover Publications Inc., 1954)
12. E. Rosa, F. Grover, *Formulas and Tables for the Calculation of Mutual and Self-Inductance* (US Dept. of Commerce and Labor, Bureau of Standards, Chicago, 1912)

Chapter 5
Inductive Levitation Micro-Systems

Application of inductive levitation micro-systems as micro-bearings is considered. In particular, the analysis of their stability and dynamics based on the analytical and quasi-finite element approach presented in Chaps. 3 and 4, respectively, is conducted by accompanying with experimental measurements included electrical and physical parameters of the system such as coil impedance, levitation height and operating temperature of micro-coils. Also, the inductive micro-bearing with the lowest energy dissipation is discussed.

5.1 Micro-Bearings

A schematic of 3D micro-bearing system is shown in Fig. 5.2d. It mainly consists of two coaxial coils excited with ac currents which are in anti-phase to each other. Alternating currents passing through the levitation and stabilization coils create a magnetic flux. This flux intercepts the conducting disc and induces eddy current, which in turn generates a magnetic field to produce a repulsive force that levitates stably the disc. The role of the inner coil (levitation coil) is to levitate a conductive disc placed on top of the coil structure, while the role of the outer coil (stabilization coil) is to provide the lateral stabilization.

5.1.1 Design and Fabrication

The micro-fabricated structure of 3D coils is presented in Fig. 2.3a. An insulating substrate (glass or Pyrex) is metallized and pads for electrical contacts are defined by standard UV photolithography, electroplating and wet chemical etching as dis-

K. Poletkin, *Levitation Micro-Systems*, Microsystems and Nanosystems, https://doi.org/10.1007/978-3-030-58908-0_5

cussed in Chap. 2. In the particular prototype shown in Fig. 2.3a, contact pads were electroplated with a thickness of 10 μm , to support better reliability for the wire-bonding process, and to decrease contact resistance. Then, an SU-8 2150 layer of 700 μm thick was cast on the wafer and structured by UV lithography to define the cylindrical pillars for subsequent wire-bonding of the coils. Thicker sidewalls of the cylinders or even full pillars are beneficial for the adhesion of SU-8 structures on the Pyrex substrates with ensuring increased stiction between the pillars. In particular, the sidewalls of two SU-8 cylinders for the outer and inner coils were fabricated to be 200 μm. In the last step, the coils were manufactured using an automatic wire-bonder which allows us to freely define the total number of windings per coil, the pitch between the windings and the number of winding layers.

Adding winding layers, a number of windings in a 3D coil can be increased drastically. In turn, it leads to an increase in a number of ampere-turns per coil still using the same height of the coil. However, it comes with a couple of inherent drawbacks. The amount of heat generated in the same volume increases with the number of layers, i.e., the number of overlapping windings, eventually leading to damaging the wire insulation, short-circuit and failure of the device. The second limitation brought along by multiple layers of windings concerns the operation range in terms of frequency. A higher operating frequency corresponds to a higher variation speed of the magnetic flux, consequently larger value of induced eddy current under the same magnitude of the excitation current in the coils. However, multiple layers of winding decrease the self-resonant frequency of the corresponding coil, therefore the border where the coil does not behave as an inductor any longer. To compromise these two mentioned aspects, one single layer of windings for both levitation and stabilization coils was used. The levitation coil was formed by 20 windings, while the stabilization coil by 12 windings. Laser cutting using Trumpf TruMark Station 5000 was employed to fabricate the disc-shaped proof mass. Micro-discs were cut from a 25 μm-thick aluminium foil (Advent Research Materials) glued to a glass wafer by thick AZ9260 photoresist. In this particular case, the aluminium micro-disc was cut to a diameter of 3200 μm. All the parameters of the 3D micro-bearing system used are summarized in Table 5.1.

Table 5.1 Parameters of the 3D prototype of the micro-bearing system

Radius of the levitation coil, r_l	1000 μm
Radius of the stabilization coil, r_s	1900 μm
The coils pitch of winding, p	25 μm
Number of windings for stabilization coil, N	12
Number of windings for levitation coil, M	20
Radius of micro-disc, r_{pm}	1600 μm
Thickness of micro-disc	25 μm

5.1.2 Measurement of Stiffness

In this section, an experimental approach to analyze static and dynamic properties the 3D micro-bearing system is discussed. The approach is based on the measurement of the stiffness of the system relative to the appropriate generalized coordinates q_v, q_l and θ according to mathematical model corresponding to the axially symmetric structure (see Table 3.1). The stiffness components can be evaluated by direct measurements of an applied force to a particular point of the levitated micro-disc simultaneously with its linear displacement. Such measurements can be performed, for instance, by using a mechanical probe FT-FS1000 developed by FEMTO-TOOLS company. Note that the probe supports the measurement of the force and linear displacement with a resolution of 0.005 μN and 5 nm, respectively. The experimental setup and measurements are described below in detail.

Experiment Setup

Figure 5.1a shows the scheme of the experimental setup. Each coil was fed with a square wave ac current provided by a current amplifier (LCF A093R). The amplitude and the frequency of the current in each coil were controlled by a function generator (Arbstudio 1104D) via a computer. To prevent the collision of the sensing probe tip having a thickness of 50 μm with SU-8 pillars or the coils, the micro-disc must be levitated at least at a height of 50 μm measured from the top of the SU-8 pillars. To fulfil this condition, a levitation height using a laser distance sensor (LK-G32) was calibrated. As a result, the measured root mean square (rms) current in the stabilization and levitation coils were to be 0.106 A at a frequency of 12 MHz, which levitated the disc at height of 54 μm measured from the top of the SU-8 pillars. Taking into account that the last coil winding for this particular structure ends 60 μm away from the top of the SU-8 structure, the total actual levitation height, h, was 104 μm. Once the calibration of the levitation conditions had been performed. Then, the micro-force sensing probe was moved towards the micro-disc until mechanical contact between the disc and probe occurred, as shown in Fig. 5.1b. Now the setup is ready to start the measurement. During the measurement, the mechanical probe pushed the disc out of the initial equilibrium state. The applied force to the disc and the linear displacement along the direction of action of this force were recorded.

Measurement

For the evaluation of the stiffness component in the radial direction relative to the generalized coordinate q_l, the force must be applied to the edge of the disc. The sensing probe was tilted by an angle of 15° with respect to the plane of the disc. The result of this measurement is shown in Fig. 5.2. The stiffness in the radial direction, k_l, was calculated to be 3.0×10^{-3} N $\cdot$ m^{-1}. For this calculation, we chose a measurement range up to 200 μm, exhibiting a linear dependence of the applied force on the linear displacement, as shown in Fig. 5.2.

However, in this experimental setup, only the force and linear displacement can be measured directly and, due to this fact the angular stiffness, k_θ, relative to the generalized coordinate, θ, cannot be evaluated from one single measurement. Also,

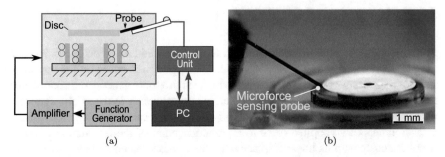

(a)   (b)

Fig. 5.1 a Scheme of the experiment setup for mechanical property measurement. A USB microscope was employed to capture probe motions. The 3D micro-bearing and probe were not shown in scale. **b** Picture of the micro-force sensing probe pushing a disc in the lateral direction. The sensing probe was tilted by 15° (this is a link on YouTube video: https://youtu.be/hLaEyYwNCs0)

Fig. 5.2 Lateral force and displacement of the probe in the horizontal plane. The angle between horizontal plane and probe was 15°. The movement range was 370 μm

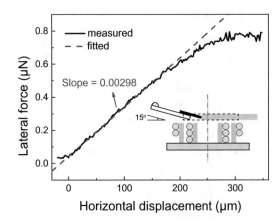

to evaluate the vertical stiffness, k_v, relative to the generalized coordinate, q_v, the probe should be applied exactly at the point of the resultant force acting on the disc in the vertical direction. However, the force centre does not coincide with the geometric centre of the plate surface of the disc, which makes very problematic applying the tip exactly to the force centre.

To avoid the difficulties mentioned above, k_θ and k_v were evaluated by using an indirect method. The measurements were performed at two different points located at $r_{p1} = 300$ μm and $r_{p2} = 1450$ μm from the disc centre. The results of these measurements are shown in Fig. 5.3a, b. In the particular points chosen for these measurements, the following stiffness values were calculated $k_{p1} = 3.5 \times 10^{-2}$ N · m^{-1} and $k_{p2} = 6.0 \times 10^{-3}$ N · m^{-1} for r_{p1} and r_{p2}, respectively. Substituting k_{p1}, k_{p2}, r_{p1} and r_{p2} into the set below:

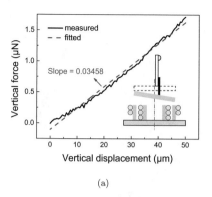

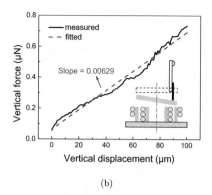

(a)   (b)

Fig. 5.3 Vertical force and displacement of the probe (this is a link on Youtube video: https://youtu.be/4EiR_k5Ouuo) : **a** The probe initial position 300 μm from the centre of the disc. The movement range was 50 μm. **b** The probe initial position 1450 μm from the centre of the disc. The movement range was 100 μm

$$\begin{cases} k_\theta = \dfrac{\left(r_{p2}^2 - r_{p1}^2\right) k_{p1}}{k_{p1}/k_{p2} - 1}; \\ k_v = \dfrac{k_{p2}}{1 - r_{p2}^2 \left(k_{p2}/k_\varphi\right)}, \end{cases} \tag{5.1}$$

the desired component of stiffness can be calculated. Hence, we have the vertical stiffness k_v, relative to the generalized coordinated, q_v: $k_v = 4.5 \times 10^{-2}$ N · m^{-1} and the angular stiffness k_θ, relative to the generalized coordinate, θ: $k_\theta = 1.5 \times 10^{-8}$ N · m · rad^{-1}.

5.1.3 Modelling

In this section, the 3D micro-bearing system shown in Fig. 3.2d is analysed theoretically by employing the analytical approach presented above. As it was shown in work [1] and it will be discussed in Sect. 6.1.3 below, the induced eddy current is distributed along with the levitated disc in such a way that two circuits having maximum values of eddy current density can be identified. Hence, the eddy current circuit can be represented as shown in Fig. 5.4a. The eddy current i_1 flows along the edge of the disc. At the same time, the circuit for eddy current i_2 is defined by the levitation coil and has a circular path with a radius equal to the radius of the levitation coil. Unlike the i_1 circuit, the position in space of the i_2 circuit is dependent only on the two generalized coordinates θ and q_v, and independent on the lateral displacement q_l of the disc, as shown in Fig. 5.4b. This figure presents the case for which the lateral displacement of the disc takes place along the Y' axis.

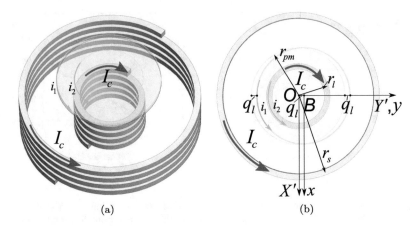

(a) (b)

Fig. 5.4 Schematic of 3D micromachined inductive micro-bearing with two representative circuits for the induced eddy current

Now let us assume that the levitation micro-bearing is operated in air. Hence, the elements of the matrix $\underline{R} = (\mathrm{Re}\{\overline{c}_{lr}\})$ will be defined as functions of the design parameters. Thus, $d = r_{pm} - r_l$ is the difference between the radius of the disc and the radius of the levitation coil (see, Fig. 5.4b), and h is the levitation height. To calculate the stiffness components as described by (3.41), Eq. (3.39) are compiled as follows.

To estimate stiffness components for the model of 3D axially symmetric inductive levitation micro-bearing system, the structure of which has defined in Table 3.1, terms of Taylor series in (3.39) are calculated. Using the simplified formula (4.33) for calculation of mutual inductance described in Sect. 4.3.2 and Fig. 5.4b, we can write for m_0^{kj}:

$$
\begin{aligned}
m_0^{11} &= \sum_{\iota=0}^{N-1} \mu_0 \cdot r_s \left[\ln \frac{8 r_s}{\sqrt{(h + \iota \cdot p)^2 + (d - c)^2}} - 1.92 \right]; \\
m_0^{22} &= \sum_{\iota=0}^{M-1} \mu_0 \cdot r_l \left[\ln \frac{8 r_l}{h + \iota \cdot p} - 1.92 \right]; \\
m_0^{12} &= \sum_{\iota=0}^{M-1} \mu_0 \cdot (r_l + d) \left[\ln \frac{8 (r_l + d)}{\sqrt{(h + \iota \cdot p)^2 + d^2}} - 1.92 \right]; \\
m_0^{21} &= \sum_{\iota=0}^{N-1} \mu_0 \cdot r_s \left[\ln \frac{8 r_s}{\sqrt{(h + \iota \cdot p)^2 + c^2}} - 1.92 \right],
\end{aligned}
\tag{5.2}
$$

where $\mu_0 = 4\pi \times 10^{-7}$ H/m is the magnetic permeability of vacuum, p is the winding pitch of the coils, $c = r_s - r_l$, N and M are numbers of winding for stabilization and levitation coils, respectively. According to [2] for the set of terms m_l^{kj} we have

$$m_\nu^{11} = -\sum_{\iota=0}^{N-1} \mu_0 \cdot r_s \frac{h + \iota \cdot p}{(h + \iota \cdot p)^2 + (c - d)^2};$$

$$m_\nu^{22} = -\sum_{\iota=0}^{M-1} \mu_0 \cdot r_l \frac{1}{h + \iota \cdot p};$$

$$m_\nu^{12} = -\sum_{\iota=0}^{M-1} \mu_0 \cdot (r_l + d) \frac{h + \iota \cdot p}{(h + \iota \cdot p)^2 + d^2}; \qquad (5.3)$$

$$m_\nu^{21} = -\sum_{\iota=0}^{N-1} \mu_0 \cdot r_s \frac{h + \iota \cdot p}{(h + \iota \cdot p)^2 + c^2};$$

$$m_l^{11} = m_l^{22} = m_l^{12} = m_l^{21} = 0;$$
$$m_\theta^{11} = m_\theta^{22} = m_\theta^{12} = m_\theta^{21} = 0.$$

Terms of the second derivatives, m_{ll}^{kj}, are defined as follows. For $m_{\nu\nu}^{kj}$, we have:

$$m_{\nu\nu}^{11} = \sum_{\iota=0}^{N-1} \mu_0 \cdot r_s \frac{(h + \iota \cdot p)^2 - (c - d)^2}{\left[(h + \iota \cdot p)^2 + (c - d)^2\right]^2};$$

$$m_{\nu\nu}^{22} = \sum_{\iota=0}^{N-1} \mu_0 \cdot r_l \frac{1}{h + \iota \cdot p};$$

$$m_{\nu\nu}^{12} = \sum_{\iota=0}^{N-1} \mu_0 \cdot (r_l + d) \frac{(h + \iota \cdot p)^2 - d^2}{\left[(h + \iota \cdot p)^2 + d^2\right]^2}; \qquad (5.4)$$

$$m_{\nu\nu}^{21} = \sum_{\iota=0}^{N-1} \mu_0 \cdot r_s \frac{(h + \iota \cdot p)^2 - c^2}{\left[(h + \iota \cdot p)^2 + c^2\right]^2}.$$

Taking into account the behaviour of the second circuit of eddy current shown in Fig. 5.4b, for m_{ll}^{kj}, we can write:

$$m_{ll}^{11} = \sum_{\iota=0}^{N-1} \mu_0 \cdot \frac{r_s}{2} \frac{(d - c)^2 r_s - (h + \iota \cdot p)^2 (r_s - 2c + 2d)}{(r_s - (c - d)) \left[(h + \iota \cdot p)^2 + (d - c)^2\right]^2};$$

$$m_{ll}^{12} = \sum_{\iota=0}^{N-1} \mu_0 \cdot \frac{r_l}{2} \frac{d^2 r_l - (h + \iota \cdot p)^2 (r_l + 2d)}{(r_l + d) \left[(h + \iota \cdot p)^2 + d^2\right]^2} \qquad (5.5)$$

$$m_{ll}^{21} = m_{ll}^{22} = 0.$$

For $m_{\theta\theta}^{kj}$, we can write:

$$m_{\theta\theta}^{11} = (r_l + d)^2 \sum_{\iota=0}^{N-1} \mu_0 \cdot \frac{r_s}{2} \frac{(h + \iota \cdot p)^2 - (d - c)^2}{\left[(h + \iota \cdot p)^2 + (d - c)^2\right]^2};$$

$$m_{\theta\theta}^{12} = r_l^2 \sum_{\iota=0}^{N-1} \mu_0 \cdot \frac{r_s}{2} \frac{(h + \iota \cdot p)^2 - d^2}{\left[(h + \iota \cdot p)^2 + d^2\right]^2}; \qquad (5.6)$$

$$m_{\theta\theta}^{21} = (r_l + d)^2 \sum_{\iota=0}^{M-1} \mu_0 \cdot \frac{r_l}{2} \frac{(h + \iota \cdot p)^2 - d^2}{\left[(h + \iota \cdot p)^2 + d^2\right]^2};$$

$$m_{\theta\theta}^{22} = r_l^2 \sum_{\iota=0}^{M-1} \mu_0 \cdot \frac{r_l}{2} \frac{1}{(h + \iota \cdot p)^2}.$$

Terms of Taylor series for self and mutual inductances of eddy current circuits like L_{ks}^0 can be written as follows. Terms L_{ks}^0 are

$$L_{11}^0 = \mu_0 \cdot (r_l + d) \left[\ln \frac{8(r_l + d)}{\delta} - 1.92\right];$$

$$L_{22}^0 = \mu_0 \cdot r_l \left[\ln \frac{8r_l}{\delta} - 1.92\right]; \tag{5.7}$$

$$L_{12}^0 = L_{12}^0 = \mu_0 \cdot (r_l + d) \left[\ln \frac{8(r_l + d)}{d} - 1.92\right],$$

where δ is the characteristic length for eddy current circuit. The characteristic length is introduced to account for the effect of frequency, which will be discussed in detail in Sect. 5.1.5 below. Using the equations above, we can define determinants (3.29) and (3.33) as follows. Determinants Δ_0^{ks} are

$$\Delta_{11} = \begin{vmatrix} m_0^{11} & m_0^{21} \\ L_{21}^0 & L_{22}^0 \end{vmatrix}; \; \Delta_{12} = \begin{vmatrix} m_0^{12} & m_0^{22} \\ L_{21}^0 & L_{22}^0 \end{vmatrix};$$

$$\Delta_{21} = \begin{vmatrix} L_{11}^0 & L_{12}^0 \\ m_0^{11} & m_0^{21} \end{vmatrix}; \; \Delta_{22} = \begin{vmatrix} L_{11}^0 & L_{12}^0 \\ m_0^{12} & m_0^{22} \end{vmatrix}. \tag{5.8}$$

Determinant Δ is

$$\Delta = \begin{vmatrix} L_{11}^0 & L_{12}^0 \\ L_{21}^0 & L_{22}^0 \end{vmatrix}. \tag{5.9}$$

We apply the obtained model to the design of the micro-bearing system considered above, namely, the radii of the stabilization and levitation coils are 1.9 and 1.0 mm, respectively; the pitch of coil winding is 25 μm; the number of windings for the stabilization and levitation coils are 12 and 20, respectively; the radius of the disc is 1.6 mm. For an excitation current of 109 mA in both coils and characteristic length δ at a frequency of 10 MHz to be 15 μm (see Table 5.3 below in Sect. 5.1.5), the maps of stiffness coefficients of the micro-bearing system are shown in Fig. 5.5. The results of the calculation are shown in Table 5.2 together with experimental and modelling results. The analysis of Table 5.2 shows that the developed technique allows us to evaluate the stiffness with a good enough accuracy. This fact proves the efficiency of the proposed analytical approach presented in Chap. 3.

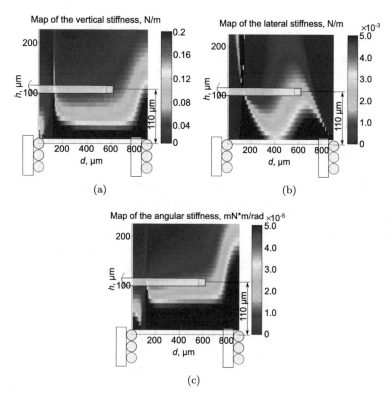

Fig. 5.5 Maps of vertical, lateral and angular stiffness coefficients of the suspension; the square with dashed lines is the area of interest for calculation

Table 5.2 Comparison of results of modelling and experiment measuement for stiffness components

Stiffness component	Measurement	Modelling
Lateral, [N · m^{-1}]	3.0×10^{-3}	2.0×10^{-3}
Vertical, [N· m^{-1}]	4.5×10^{-2}	4.5×10^{-2}
Angular, [m · N · rad^{-1}]	1.5×10^{-8}	1.4×10^{-8}

5.1.4 Coil Impedance

The major advantage of 3D wire-bonded coils compared with planar coils is the easiness of obtaining a relatively large number of windings, thus a significantly larger ampere-turn value. To achieve a larger upward vertical component of the Lorentz force, therefore a larger levitation effect, the number of windings of the inner (levitation) coil, should be increased. The parameters of the outer (stabilization) coil, such as position, diameter, number of windings and current intensity contribute to the stabilization of the levitated disc and are reflected in the stiffness of the structure (please see also work [3]). The wire-bonding process used in the fabrication of the 3D solenoidal coils has been explained in the previous chapter (see Chap. 2). Basically, the movement of the head of an automatic wire-bonder is controlled via a MATLAB interface where a 3D helical trajectory is defined. The head of the wire-bonder moves according to the defined trajectory around a pillar structured in thick SU-8 using UV lithography. The wire plastically deforms to the shape of the SU-8 pillar remaining as a solenoidal 3D coil. The typical height of the SU-8 pillars was 650 μm, and there is no pitch between adjacent wires. The wire considered here is insulated gold wire, 25 μm in diameter.

Also, the application of 3D coils for micro-bearing systems requires to take into account their electrical impedance. Firstly, the electrical impedance of the 3D coils influences on the operation frequency of the device, which should be well below the resonant frequency of the coils. When the frequency of excitation currents gets close to the resonant frequency, the coil impedance increases dramatically, making the bearing system hard to drive. Secondly, the two coils should have similar impedance values for the easiness of operation and control of the device.

The resonant frequency of the coil, f_0, is defined by

$$f_0 = \frac{1}{2\pi}\sqrt{\frac{1}{LC} - \left(\frac{R}{L}\right)^2}, \tag{5.10}$$

where L, C and R are the inductance, stray capacitance and resistance of the coil, respectively. In the case of single-layer 3D micro-coils, the inductance is very well approximated by Wheeler's formula:

$$L = \frac{10\pi \mu_0 N^2 r^2}{9r + 10h_{coil}} = \frac{10\pi \mu_0 N^2 r^2}{9r + 10Np}, \tag{5.11}$$

where μ_0 is the vacuum permeability, N is the number of windings, r is the coil radius, h_{coil} is the coil height and p is the pitch between two adjacent windings and, in our particular case, p is equal to the wire diameter. Even though 3D coils have larger h_{coil}, by rewriting h_{coil} as Np in Eq. (5.11), it can be assumed that L is proportional to N. Therefore, when employing 3D micro-coils, the inductance values are much larger than in the case of 2D micro-coils. The stray capacitance also increases significantly because the wires are adjacent to each other. As a consequence, following from Eq. (5.10), resonant frequencies of the 3D coils composing the micromachined inductive bearings are much lower compared to the 2D case. This is another important difference that needs to be taken into account when employing 3D coils with respect to their 2D counterparts.

The algorithm proposed in [4] is used for the calculation of self- and mutual inductances of coaxial current coils over the whole range of coil sizes and shapes, being fast and accurate. As the material properties and the thickness of the wire insulation layer are not disclosed due to commercial interest, we could only roughly estimate the stray capacitance of the coils using the method presented in [5], thus the

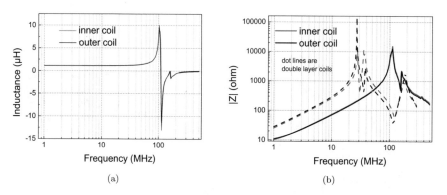

Fig. 5.6 **a** Inductance and **b** impedance measurement of coils with one layer and two layers of windings. The number of windings in the double-layer coils is twice as in the single-layer coils

resonant frequency. When the inner coil has a diameter of 2.0 mm and 20 windings arranged in one layer, the calculated self-inductance was 1.19 µH, while the stray capacitance value was 2.19 pF. Therefore, the coil inductance is the dominant factor for coil impedance. In order to obtain the same inductance, the outer coil should have 12 windings arranged in one layer for a diameter of 3.9 mm. The diameter difference between the two coils comes from the SU-8 sidewall thickness and the minimum requirement of post-capillary distance in the wire bonding process (400 µm), which determines a minimum diameter difference between the outer and inner coil. In this work, the number of winding layers was set to 1, so that a large tuning range of both input current amplitude and frequency was possible for experimental characterization purposes. The detailed fabrication process can also be found in [3].

In order to verify our assumptions and the theoretical calculations and to determine the device working frequency range, the impedances of inner and outer coils using an impedance analyser, Agilent 4991A were measured. The results are shown in Fig. 5.6. The inductance of the single-layer coil far left from the resonant frequency was measured to be 1.16 µH, with the resonant frequency of 111 MHz. The inductance of the 2D micro-bearing system in [6] with the same size in diameter was only about 10 nH and a resonant frequency of 15 GHz according to simulations carried out with Agilent Advanced Design Systems. The repeatability of coil electrical parameters for single-layer coils was reported to be excellent in [7].

For comparison, Fig. 5.6b also includes the impedance of coils having twice the number of windings arranged in two layers, showing that the resonant frequency for these structures drops almost four times compared to the one-layer structures. This proves that the potential working frequency range is further drastically reduced in double-layer coils.

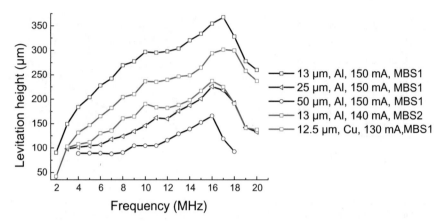

Fig. 5.7 Levitation height as a function of input current frequency

5.1.5   Levitation Height

The dependency of the levitation height on the input current frequency for discs with various thicknesses, 13, 25 and 50 μm for aluminium and 12.5 μm for copper, and, therefore, different ratios of skin depth to thickness is studied. These measurements have been performed with a 3D micro-bearing structure having the inner (levitation) coil with a diameter of 2.0 mm and 20 windings and the outer (stabilization) coil with a diameter of 3.9 mm and 12 windings, which is denoted as MBS1 (Micro-Bearing System) in Fig. 5.7. Additionally, we studied another 3D micro-bearing structure having a different number of windings for the inner (15 windings) and outer (9 windings) coils, respectively (MBS2). All measurements have been performed using a square wave excitation current in both inner and outer coils. The rms values for the currents used in each experiment are indicated in Fig. 5.7.

Figure 5.7 summarizes the experimental results showing the levitation height as a function of current frequency for constant rms current values. From 2 to 8 MHz, the slopes of the curves corresponding to the disc with a thickness of 13 μm are steeper than the slopes of the curves corresponding to thicker discs. Similarly, the slopes of the curves corresponding to 13 μm Al disc are steeper than the curve for 12.5 μm-thick Cu disc. This verifies the expected effect of the skin depth, which is more obvious for thicker materials, as well as for materials with higher conductivities. At 4 MHz, the skin depth for aluminium is 41 μm and for copper is 32.6 μm, which means that the thinner discs (12.5 and 13 μm) do not yet feel the skin effect, while this is already present for the thickest (50 μm) disc. A very important remark for these experiments is the fact that at higher frequencies, the square wave signal generated by the function generator gets distorted by the real characteristics of the circuit. A square wave signal of frequency f can be seen as a superposition of sine wave signals with frequency multiples of f, which will be amplified differently according to the gain versus frequency curve of the amplifier. Therefore, at higher nominal frequencies

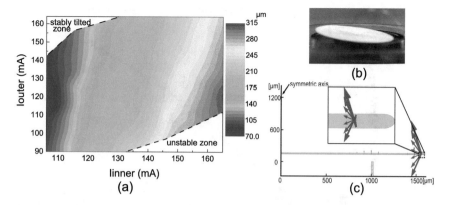

Fig. 5.8 **a** Levitation height as a function of rms values of input currents; **b** Photo of the tilted disc; **c** Lorentz force distribution (red arrow) under the condition of $I_{outer} = 2I_{inner} = 220$ mA

of the input square wave, one will notice a mixture of phenomena arising from the fact that the coils composing the micro-bearing structure are actually experiencing frequencies much closer to resonance, as well as the mutual coupling between coils and a disc. For the considered designs of levitation micro-bearing system, the skin effect decreases the levitation height at a frequency above 18 MHz.

The use of two amplifiers allows the independent study of the effects of the inner and outer coils. Figure 5.8 provides a map of the levitation height as a function of the inner coil and outer coil current rms values. The levitation height is increased with the increase of the inner (levitation) coil current and is decreased with the increase of the outer (stabilization) coil current. This fact agrees with the general operation principle: the inner coil mainly contributes to the levitation effect, while the outer coil is responsible for the system stability. By referring to Fig. 5.9, we find that the micro-bearing showed a good linear control of levitation height with respect to coil current. When the bearing is biased in the region characterized by large inner coil current and low outer coil current, the disc cannot be levitated stably, the so-called "unstable zone". When the current in the outer coil is much larger than in the inner coil, the disc is stably tilted, as shown in Fig. 5.8b. Moreover, COMSOL simulation of this special bias condition shows that the Lorentz force is focused at the edge of the disc. This tilting of disc is due to the fact that the real micro-bearing structure is asymmetric because of the spiral nature of the coils and the fact that the disc may not be perfectly flat and symmetrical. Any of these small asymmetries leads to the so-called "stably-tilted zone".

In order to compare the levitation effects of 3D to 2D micro-bearing system, we scaled up the 2D micro-bearing system reported by Shearwood in [6] to the same diameter of our 3D micro-bearing system. Using this scaled-up 2D micro-bearing system, COMSOL simulation shows that it needs 12 times more input current compared to our 3D micro-bearing system (140 mA at 8 MHz to levitate the same disc at the same height.

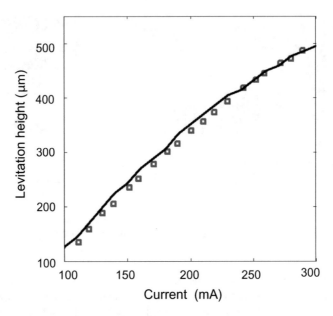

Fig. 5.9 The red square points are experimental measurement. The black line is the theoretical curve generated by Eq. (5.14) assuming that $\delta = 20\,\mu$m and $\alpha = 0.8$

Theoretical Calculation of the Input Current and Frequency

Let us consider the equilibrium state of the micromachined inductive bearing in the vertical direction. As the simplest case, one eddy current circuit approximation is used to describe the equilibrium state. Hence, we can write [8]:

$$mg + \frac{I^2}{L^o} \frac{\partial M}{\partial q_v}\bigg|_{q_v=h} M = 0, \tag{5.12}$$

where I is the current in the coils (assuming the same current in both coils), L^o is the self-induction of the disc and M is the mutual induction between the coils and the disc; in this particular case, the mutual induction can be considered as a function of the disc vertical displacement characterizing generalized coordinate q_v. The equilibrium occurs at the point $q_v = h$. The second term in Eq. (5.12) represents the Ampère force acting between the coil and eddy currents and obtained as the derivative of the potential energy with respect to the vertical displacement.

The numerical analysis (see Chap. 6) shows that for a particular micro-bearing design, the contribution of the levitation coil in the levitation force is significantly larger compared to the contribution of the stabilization coil. Additionally, the levitation coil induces the eddy current in the disc, the maximum value of which is distributed within a ring with the same radius as the levitation coil, as shown in [1]. Assuming that the levitation height of the disc is small and the fact that the function of mutual inductance between the disc and the ring-shaped coils can be simplified and expressed in terms of simple functions, the qualitative approach developed in [2] can be applied to the present study. Hence, the mutual induction can be represented in quadratic form as follows:

$$M = m_0 + m_1 q_v + m_2 q_v^2, \tag{5.13}$$

where m_0, m_1 and m_2 are the coefficients of the quadratic form defined in equilibrium position ($q_v = h$). Static Eq. (5.12) can be rewritten in terms of the coefficients of this quadratic form as follows:

$$mg - \frac{I^2 m_1(h/\alpha(\omega)) m_0(h/\alpha(\omega))}{L^o} = 0, \tag{5.14}$$

where α is an implicit function that takes into account the scaling effect of the excitation current frequency, ω, i.e, the increase of the maximum value of the eddy current with the increase of frequency. According to [3], the coefficients $m_1(h)$ and $m_0(h)$ can be defined as follows:

$$m_0(h) = \sum_{v=0}^{N-1} r_l \mu_0 \left[\ln \frac{8 r_l}{(h + v \cdot p)} - 1.92 \right]; \; m_l(h) = \sum_{v=0}^{N-1} \frac{r_l \mu_0}{(h + v \cdot p)}, \tag{5.15}$$

where N is the number of turns of the levitation coil, r_l is the radius of the levitation coil and p is the pitch between consecutive windings of the coil. The self-induction of the ring-shaped path of the eddy current is defined as follows:

$$L^o = r_l \mu_0 \left[\ln \frac{8 r_l}{\delta(\omega)} - 0.3 \right], \tag{5.16}$$

where δ is a frequency-dependent function whose physical meaning is the effective width in which the maximum of the induced eddy current within the disc is distributed. This dependency is a result of changing the sharpness of the distribution of the eddy current with frequency. As shown in [1], the effective width can be evaluated as $\delta = 0.01..0.1 \cdot r_l$.

Equation (5.14), together with the parameters defined by $\delta(\omega)$ and $\alpha(\omega)$, can be used to completely determine the state of the device, i.e., levitation height, input current and frequency. Figure 5.9 shows that a set of experimental data points (the red square points) representing levitation height versus excitation current in the levitation coil can be fitted with the theoretical curve generated by Eq. (5.14) (the black line) for two properly chosen values for $\delta(\omega)$ and $\alpha(\omega)$. The measurement was carried out for a disc with a radius of 1.6 mm and a thickness of 13 μm, which was levitated by a levitation coil with a radius of 1 mm and 15 windings, fed by ac with a frequency of 10 MHz. The fitting by Eq. (5.14) was performed assuming that $\delta = 20$ μm and $\alpha = 0.80$.

Table 5.3 Defined α and δ for four frequency reference points

Frequency, MHz	3	6	8	12
α	0.59	0.7	0.77	0.84
δ, μm	45	25	18	14

Both functions, namely, $\delta(\omega)$ and $\alpha(\omega)$, can be defined experimentally for a particular range of frequencies in a manner similar to the measurement and fitting presented in Fig. 5.9. The measurements should be conducted for different reference points within the desired range of frequencies. For instance, in the case of a disc with a thickness of 13 μm and a levitation coil with a radius of 1 mm and 20 windings, we define the curves of $\alpha(\omega)$ and $\delta(\omega)$ within the frequency range from 2 to 14 MHz. We used four arbitrary reference points at 3, 6, 8 and 12 MHz. For these reference points, parameters α_n and δ_n ($n = 1$–4) are extracted from experimental measurements as explained above and listed in Table 5.3. Using Table 5.3, the behaviour of both $\alpha(\omega)$ and $\delta(\omega)$ within the chosen frequency range can be fitted by polynomial functions, as shown in Fig. 5.10.

Figure 5.10 confirms that increasing the frequency of the excitation current, the eddy current distribution becomes more confined, i.e., the α parameter gets narrower, while the maximum of the eddy current, i.e., the δ parameter, increases.

To verify this result obtained from basic theoretical considerations, the current needed to achieve a certain levitation height is estimated according to the formula derived from Eq. (5.14):

$$I = \sqrt{\frac{mg L^0}{m_1(h/\alpha(\omega))m_0(h/\alpha(\omega))}}. \tag{5.17}$$

The theoretical predicted curve is displayed in Fig. 5.11 (the solid black line). For comparison, we have plotted on the same graph the experimental measurement data points (the red circles) for the following frequencies, 3, 4, 5, 6, 8, 9 and 10 MHz, showing a good agreement with the theoretical curve. This result confirms the qualitative prediction that increasing the frequency, a lower current amplitude is necessary in order to achieve the same levitation effect, but at the same time, represents a quantitative prediction method of the current needed for a certain levitation height at a certain frequency of the excitation signal. At the same time, the qualitative Eq. (5.14) can be expanded to a wider frequency range by adapting the coefficients of current amplitude scaling, $\alpha(\omega)$, and the effective width, $\delta(\omega)$.

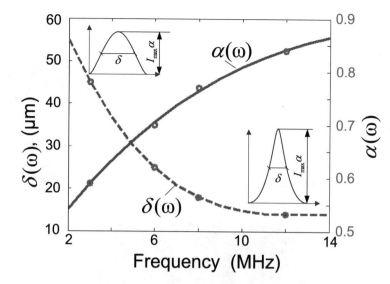

Fig. 5.10 Curves of $\alpha(\omega)$ (blue line) and $\delta(\omega)$ (dashed line), together with α_n and δ_n ($n = 1$–4) (red circles) taken from Table 5.3

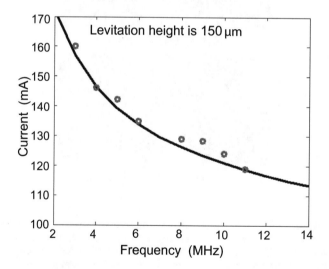

Fig. 5.11 Current versus frequency dependence for constant levitation height

5.1.6 Lateral Stability

A proper characterization of the disc stability is important for any potential application of micro-bearing levitation systems. In particular, the *lateral stability* is experimentally investigated by means of measuring lateral force F_r in the lateral displace-

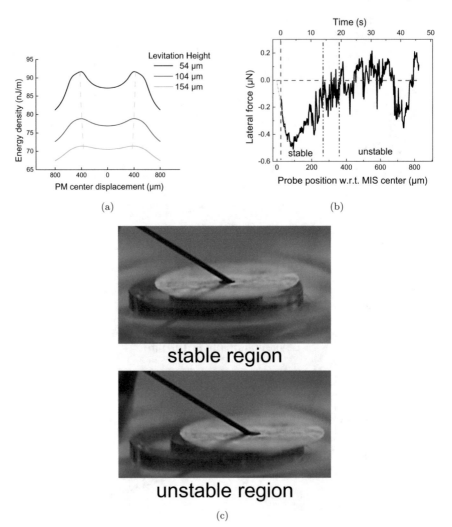

Fig. 5.12 **a** Measurement of force versus displacement. The region between the vertical blue lines represents the transition between the two states depicted in **b**; **b** captures of the lateral displacement measurement corresponding to a stable (top) and unstable (bottom) position, respectively; **c** The system energy wells obtained by 2D simulation with variable disc centre position

ment of the disc with a mechanical probe (FT-FS1000, FEMTO-TOOLS AG Switzer-land) and a micro-force sensing probe (FT-S100, FEMTO-TOOLS AG Switzerland) with a force and linear displacement resolution of 5 nN and 5 nm, respectively. The mechanical probe was placed under a 30° with respect to the plane of the disc and inserted into a hole with a diameter of 400 μm, cut in the middle of the disc, as depicted in Fig. 5.12b. The probe is able to move at a constant step speed and record the force together with the displacement. The force versus displacement curve is presented in Fig. 5.12a. The analysis of Fig. 5.12a shows that the behaviour of the lateral force can be defined in three regions of lateral displacement, namely, stable, transition and unstable. In the stable region from 0 to 260 μm, the lateral force is a restoring force, pushing the disc back to the equilibrium position. In the transition region defined by the two blue dashed lines, F_r approaches zero. In the unstable region, F_r is a runaway force, pushing the disc away from the equilibrium position.

The lateral stability can also be investigated from an energetic point of view. The lateral force and system energy can be linked according to the definition:

$$F_l = -\frac{\partial W_m}{\partial q_l}, \tag{5.18}$$

where W_m and q_l are the system energy and the displacement in the radial direction, respectively. The profile of the system energy has been obtained by 2D COMSOL simulation (Fig. 5.12c). The exact transition point between the stable and unstable region is where the energy reaches the maximum and its derivative F_l becomes zero. The difference between simulation and experiment should have been due to the 2D model, lacking disc velocity and collisions. It is worth noting that the disc is less stable (the potential energy well is shallower) and less confined (the stability region is wider) for higher levitation heights.

5.1.7 Temperature

Another major issue in using the electromagnetic induction in micro-devices is the large current density required due to scaling down, thus leading to increased heat dissipation and temperature in micro-devices, which may cause melting, metal delamination, oxidation and, eventually, complete device failure. For the 2D micro-bearing system developed by Shearwood's group [6], the temperature of the coil was reported to be 600 °C. Because 3D micro-coils offer the possibility to employ lower currents, it also seems more competitive from the point of view of the heat dissipated in the device volume. On the other hand, the higher density of windings of the 3D micro-coil means that the heat is generated in a relatively small volume, and it may be difficult to efficiently evacuate the heat in order to prevent device overheating.

The temperature distribution was measured when the disc was levitated at 120 μm, as shown in Fig. 5.13. The maximum temperature was 112 °C at the outer coil. Then, by removing the disc, but maintaining the corresponding excitation of the two coils, we measured the maximum temperature of 131 °C at the inner coil, significantly lower than the 2D micro-bearing system reported by Shearwood [6].

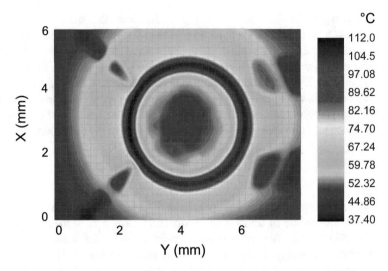

Fig. 5.13 Temperature distribution of the 3D MIS and a 3.2 mm-diameter Al PM (blue in the centre) after the MIS reached thermal equilibrium. The inner coil is blocked by the PM. IR camera resolution: 0.1 mm

5.2   Micro-Bearings with Lowest Energy Consumption

The operating temperature of inductive levitation micro-bearing systems can be further decreased by integrating 3D micro-coils with magnetic material for magnetic flux concentration and as a result, decreasing drastically the dissipation of the magnetic energy stored in 3D micro-coils. For instance, a *polymer magnetic composite* (PMC) can be used for such the purpose. A photograph of the prototype of the micro-bearing system with integrated PMC core is shown in Fig. 5.14a, demonstrating the successful levitation of a disc with a diameter of 3.2 mm and thickness of 13 μm

The fabrication of the device consists of two consecutive processes: coil structuring and core filling. The coil structure is composed of two coaxial 3D wire-bonded micro-coils, similar to those discussed above. In this prototype, the height of the coils is 650 μm and the number of windings is 20 and 12 for levitation and stabilization coils, respectively. Diameters of levitation and stabilization coils were 2.0 mmm and 3.9 mm, respectively. In the second process, the inner volume of coil structure was filled by the PMC core as shown in Fig. 5.14b. Figure 5.14c depicts a schematic of the cross-section of the coils. The core consists of a mixture of epoxies AW4510 and HW4804 (Huntsman Advanced Materials GmbH, Switzerland) and NiFeZn ferrite soft magnetic powder (CMD5005, National Magnetics Group, USA). The mixture of epoxy was injected by a dispenser (DX-250, OKI, USA) in the centre of the levitation coil, as well as in the space between the coils.

The operating principle of the prototype is the same as it was described earlier. However, the core helps to effectively use the volume of coils and to efficiently guide the magnetic field lines around the coils by minimizing their path through air. Due

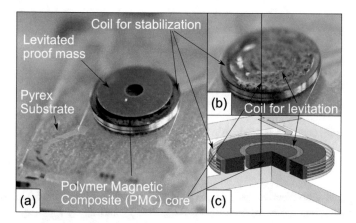

Fig. 5.14 Micromachined inductive bearing with PMC core (this is a link on YouTube video: https://youtu.be/3Z2cDCugWcU): **a** successful levitation of the disc; **b** coils with core ; **c** cross-section of coils with polymer magnetic composite core

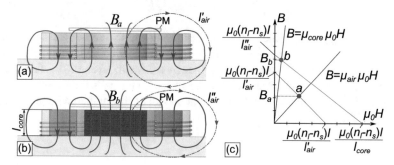

Fig. 5.15 Operating principle (blue lines and red lines correspond to the path of magnetic induction through air and core, respectively): **a** lines of magnetic induction of the bearing without core; **b** lines of magnetic induction of the bearing with core; **c** graph of estimation of magnetic inductions for the bearing with core and without one

to this fact, the magnetic induction within the area of stable levitation of the disc is increased, which results immediately on a decrease of the current required for levitation. Indeed, let us denote n_l and n_s as the number of windings for levitation and stabilization coils, respectively, and consider the inductive micro-bearing system without the core as shown in Fig. 5.15a. The magnetic field in the centre of the levitation coil without core can be estimated as $Hl'_{air} = (n_l - n_s)I$, where H is the magnetic field, l'_{air} is an effective path of magnetic field passing through the air, and I is the current through the coil. The induction, B, generated by the coil without core can be found from the graph relating B and $\mu_0 H$ as shown in Fig. 5.15c (the blue line). For this particular case, the magnetic induction is characterized by B_a. The value of B_a is the solution of the following set:

$$\begin{cases} B + \mu_0 H = \mu_0 \dfrac{(n_l - n_s)I}{l'_{air}}; \\ B = \mu_0 H, \end{cases} \tag{5.19}$$

which becomes

$$B_a = \mu_0(n_l - n_s)I/(2l'_{air}). \tag{5.20}$$

For coils with core the relationship between B and $\mu_0 H$ becomes

$$Hl_{core} + Bl''_{air}/\mu_0 = (n_l - n_s)I, \tag{5.21}$$

where l''_{air} is the effective path of magnetic field passing through the air for coils with the magnetic core. Accounting for $l_{core} < l'_{air}$ and $l''_{air} \leq l'_{air}$, Eq. (5.21) can be shown as in Fig. 5.15c (the red line). Due to the fact that μ_{core} is large [9], the magnetic induction generated by coils with core is

$$B_b \approx \mu_0(n_l - n_s)I/l''_{air}. \tag{5.22}$$

Thus, even if $l''_{air} \approx l'_{air}$ (due to the fact that the linear size of coils is one order of magnitude larger than their height), the magnetic induction generated by the coils with integrated core is a factor of 2 larger than by coils without core. This fact leads to a theoretical decrease by a factor of 2 of the current in the coils for a given levitation height, as it follows from the analysis of the model of the levitation height given in [10, p. 1478].

5.2.1   Experimental Results and Further Discussion

In order to demonstrate the positive effect of the core on the energy performance in 3D micro-bearing, the following setup was used: the coils were fed with a square wave ac provided by a current amplifier (LCF A093R); the amplitude and the frequency of the current in the coils were controlled by a function generator (Arbstudio 1104D) via a computer; the levitation height was measured by a laser distance sensor (LK-G32) having a resolution of 10 nm; the temperature of the prototype was measured by an IR (Infrared) camera PI-160 (Optris GmbH, Berlin, Germany).

Two different designs of micro-bearing systems have been fabricated and investigated experimentally: design A—only the levitation coil is filled with PMC material Fig. 5.16a, and design B—the entire volume of both coils is filled with PMC material Fig. 5.16b. The performance of these structures is compared to the performance of the structure without core. Figure 5.16c shows a micro-bearing prototype with an integrated PMC core successfully levitating an Al disc (3.2 mm in diameter and 13 μm in thickness).

The dependence between the levitation height and the excitation current in the coils at a frequency of 10 MHz has been measured experimentally and is being

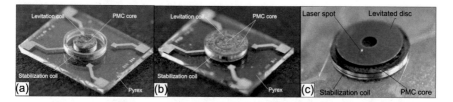

Fig. 5.16 **a** Design A: only the inner (levitation) coil is filled with PMC material; **b** design B: both the inner (levitation) and the outer (stabilization) coils are filled with PMC material; **c** an Al disc is successfully levitated on a design B

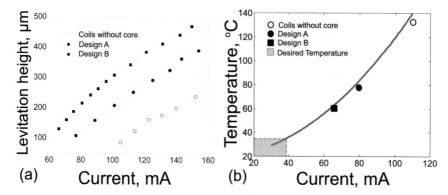

Fig. 5.17 **a** Dependence of the levitation height on the excitation current (at 10 MHz) as a function of the PMC distribution inside the MIS; **b** temperature versus current dependence for a levitation height of 110 μm

shown in Fig. 5.17a for the following micro-bearing prototypes: design A, design B and, in order to enable direct comparison, a prototype without core. The position of the three curves in Fig. 5.17a indicate that, for the same excitation current, higher levitation heights are being achieved for a higher degree of filling the coil volume with PMC material. Conversely, for the same levitation height, e.g., 110 μm, the highest actuation current is obtained for the structure without core—110 mA, and the smallest current value—70 mA, is demonstrated for design B, i.e., for the prototype having the entire coil volume filled with PMC material.

In order to check the prediction of the equations above, we choose one specific operation point for the design B prototype of the bearing: for an excitation current of 60 mA, the levitation height of the disc was 110 μm. The same levitation height (110 μm) in the micro-bearing without core was achieved by the excitation current of 110 mA, which is a factor of 1.8 higher. Equations (5.20) and (5.22) above also predict that the magnetic induction doubles for a prototype with a magnetic core, therefore, we can conclude a relatively good agreement between the theoretical prediction and the experimental results. It means that managing the distribution of magnetic field within coils by designing magnetic core the required current for levitation can be reduced together with operating temperature. Figure 5.17b sums up this fact

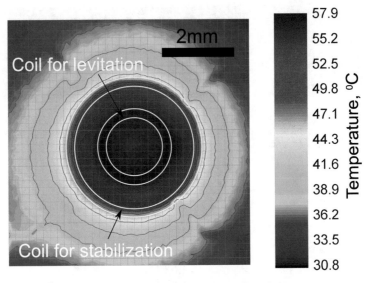

Fig. 5.18 Temperature map measured on the surface of coils for a levitation height of 110 μm and a current of 65 mA. White lines show schematically the location of the edges of the coils

by showing the dependence of operating temperature for considered coil designs on the excitation current. Figure 5.17b predicts that operating temperature can be comparable with ambient one when the excitation current will be less than 40 mA. This can be reached, for instance, by adding a PMC backing plane to the 3D micro-coils, could decrease the operating temperature to around room temperature values. A significant reduction of the current leads to a significant reduction in the Joule heating within the micro-bearing systems. This is indeed confirmed by the temperature map of the design B prototype shown in Fig. 5.18, which demonstrates that the temperature on top of the structure is lowered to 58 °C. Moreover, the temperature distribution is relatively uniform within a range of 10 °C, whereas the prototype without core shows a large temperature gradient from the coil windings to the centre of the structure [10, p. 1482].

Presented results demonstrate a synergy between 3D wire-bonding coil technology and the subsequent integration with a polymer magnetic composite material with suitable properties, i.e., high resistivity to inhibit the formation of eddy currents, and high permeability for magnetic field enhancement. Besides the record low operating current, which leads to significantly reduced i^2R power dissipation, the method offers prospects for even further improvements, one of which can be performed at the level of the magnetic core's design: by taking into account that the aspect ratio of the core, i.e., core height versus core diameter, is rather low, one can expect a higher aspect ratio core to bring an additional increase in efficiency. At the same time, the dispensing method employed to integrate the PMC core is flexible and can be used to additionally cover the bottom and outer sides of the coils, thus providing a more complete return path for the magnetic field lines (decreasing the length of l_{air}'').

References

1. Z. Lu, F. Jia, J. Korvink, U. Wallrabe, V. Badilita, Design optimization of an electromagnetic microlevitation System based on copper wirebonded coils, in *2012 Power MEMS* (Atlanta, GA, USA 2012), pp. 363–366
2. K. Poletkin, A. Chernomorsky, C. Shearwood, U. Wallrabe, A qualitative analysis of designs of micromachined electromagnetic inductive contactless suspension. Int. J. Mech. Sci. **82**, 110–121 (2014). https://doi.org/10.1016/j.ijmecsci.2014.03.013
3. Z. Lu, K. Poletkin, B. den Hartogh, U. Wallrabe, V. Badilita, 3D micro-machined inductive contactless suspension: testing and modeling. Sens. Actuators A: Phys. **220**, 134–143 (2014). https://doi.org/10.1016/j.sna.2014.09.017
4. T.H. Fawzi, P.E. Burke, The accurate computation of self and mutual inductances of circular coils. IEEE Trans. Power Appar. Syst. (2), 464–468 (1978)
5. Z. Yang, G. Wang, W. Liu, Analytical calculation of the self-resonant frequency of biomedical telemetry coils, in *2006 International Conference of the IEEE Engineering in Medicine and Biology Society* (2006), pp. 5880–5883
6. C. Shearwood, K.Y. Ho, C.B. Williams, H. Gong, Development of a levitated micromotor for application as a gyroscope. Sensor. Actuat. A-Phys. **83**(1–3), 85–92 (2000)
7. A. Moazenzadeh, N. Spengler, U. Wallrabe, High-performance, 3D-microtransformers on multilayered magnetic cores, in *2013 IEEE 26th International Conference on Micro Electro Mechanical Systems (MEMS)* (2013), pp. 287–290
8. K. Poletkin, A.I. Chernomorsky, C. Shearwood, U. Wallrabe, An analytical model of micro-machined electromagnetic inductive contactless suspension, in *the ASME 2013 International Mechanical Engineering Congress & Exposition* (ASME, San Diego, California, USA 2013), pp. V010T11A072–V010T11A072.https://doi.org/10.1115/IMECE2013-66010
9. S.G. Mariappan, A. Moazenzadeh, U. Wallrabe, Polymer magnetic composite core based micro-coils and microtransformers for very high frequency power applications. Micromachines **7**(4) (2016)
10. Z. Lu, K. Poletkin, U. Wallrabe, V. Badilita, Performance characterization of micromachined inductive suspensions based on 3D wirebonded microcoils. Micromachines **5**(4), 1469–1484 (2014)

Chapter 6
Hybrid Levitation Micro-Systems

In this chapter, applications of hybrid levitation micro-systems based on a combination of electric and inductive force fields to micro-actuators, micro-accelerators and micro-accelerometers are discussed.

6.1 Micro-Actuators

A combination of different force fields into one levitation micro-actuator allows increasing demands on its functionalities to meet emerging challenges in innovative approaches to control different physical processes occurring, for instance, in mechanics, optics and fluidics. In particular, a levitation micro-actuator combined electric and inductive force fields can execute linear and angular positioning with adjustable dynamics of the levitated micro-object, as well as operation as a linear and angular pull-in actuator. Moreover, the pull-in phenomena expand a list of the potential application of the hybrid micro-actuator. For instance, the actuator can be used as a threshold sensor [1] as well. Below, a design and micro-fabrication process based on 3D coils of such a hybrid levitation micro-actuator accompanying with its theoretical and experimental characterization of linear and angular pull-in actuation are considerd in detail.

6.1.1 Design and Micromachined Fabrication

The hybrid actuator consists of two structures fabricated independently, namely, the coil structure and the electrode structure, which are aligned and assembled by flip-chip bonding into one device with the dimensions: 9.4 mm × 7.4 mm × 1 mm as

K. Poletkin, *Levitation Micro-Systems*, Microsystems and Nanosystems, https://doi.org/10.1007/978-3-030-58908-0_6

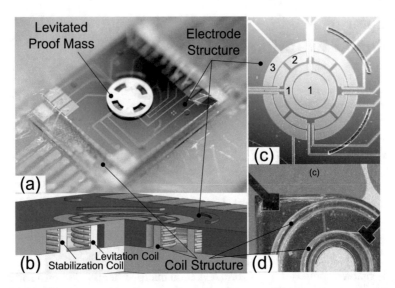

Fig. 6.1 Hybrid actuator: **a** the prototype of actuator glued to a PCB under the experimental test; **b** the scheme demonstrating the location of the coils inside the silicon structure; **c** the design of electrodes: 1-suspending, 2-rotating and 3-tilting electrodes; **d** the view of aligning electrode and coil structures from the back side (pyrex glass) of the device

shown in Fig. 6.1a. Figure 6.1b demonstrates the scheme of the actuator design with its sectional view providing the position of 3D coils inside the silicon structure. The coil structure consists of two coaxial 3D wire-bonded micro-coils similar to those considered above chapters. For this particular device, the height of the coils was 450 μm and the number of windings is 18 and 12 for levitation and stabilization coils, respectively. This coil structure is able to levitate discs with diameters ranging from 2.7 to 3.3 mm.

The electrode structure is fabricated on a 510 thick silicon substrate having a 1 μm oxide layer for passivation. Electrodes are patterned on top of the oxide layer by UV lithography on evaporated Cr/Au layers (20/150nm) as shown in Fig. 6.1c. Their functions are to achieve vertical and angular actuation of the disc. For the final assembly of the coil structure and electrode structure, the silicon substrate on which the electrodes are fabricated has been etched to a depth of 460 μm by deep reactive-ion etching (DRIE, Bosch process), aligned and bonded onto the coil structure as shown in Fig. 6.1d.

The actuator operates in the vertical direction using ac current in the coils at the frequency of 9 MHz, with a rms value ranging from 100 mA to 130 mA, which corresponds to a levitation height in a range of 35–190 μm measured from the surface of the electrodes patterned on the silicon substrate. The function of the electrode structure is three-fold: vertical actuation, as well as tilting about two horizontal orthogonal axes and oscillation about the vertical axis. By applying a voltage to the central electrodes, numbered with "1" in Fig. 6.1c, the disc is vertically actuated

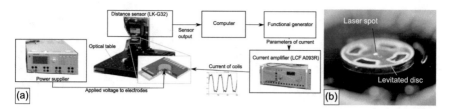

Fig. 6.2 a Experimental setup; **b** The levitated disc during the measurement of the vertical component of stiffness

upon keeping the same value of the coils current. This actuation changes the levitation height and the dynamics of the levitated disc as it will be shown below. Using the electrodes numbered with "3", the disc tilt along two orthogonal axes in the horizontal plane is controlled. The angular oscillation of the disc with respect to the vertical axis within a range of 37 degrees is reached by using the variable capacitance principle realized with electrodes numbered "2" and by carving the disc as shown in Fig. 6.1a. Note that the size and location of the four sectors, which have been cut out from the disc, are defined by the design of electrodes numbered "2" (see Fig. 6.1c).

6.1.2 Experimental Results

A scheme of the experimental setup is shown in Fig. 6.2a. Coils were fed with a square wave ac current provided by a current amplifier (LCF A093R). The amplitude and the frequency of the current in coils were controlled by a function generator (Arbstudio 1104D) via a computer. In order to characterize the actuator, its stiffness components in the vertical directions at different levitation heights were measured by recording applied voltages to appropriate electrodes and linear displacements at a point of acting a resulting electrostatic force. Linear displacements were measured by a laser distance sensor (LK-G32) having a resolution of 10 nm.

The location of the point of applying the resulted electrostatic force to the bottom surface of the disc is defined by the location and design of the electrodes. Hence, before starting the measurement, the spot of the laser sensor is located at this point without the disc. Then the disc was returned at the original place and levitated. For instance, Fig. 6.2b shows the measuring of vertical linear displacement at the central point of the disc, upon acting electrostatic force generated by energizing electrodes "1" on the bottom surface of the disc. The original equilibrium position of the disc was 120 mm, which corresponded to the rms current value of 115 mA. The pull-in effect occurred at the height of 80 mm, which agrees well with the theoretical prediction given by the quasi-FEM (please see Chap. 4 and Sect. 6.1.5). The results of measurements are shown in Fig. 6.3a in which the applied voltage, U, and the corresponding force, F, versus the linear displacement are presented. For

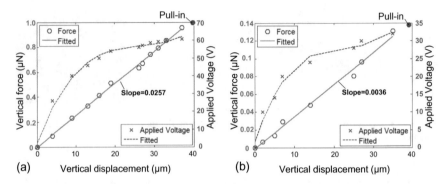

Fig. 6.3 Results of measurements for a disc with diameter of 2.8 mm: **a** for central electrodes "1" (this is a link on YouTube video: https://youtu.be/pQYzimzg4ug); **b** for outer electrodes "3" (this is a link on YouTube video: https://youtu.be/XfLMTru_vHI)

the electrodes having the same areas, A, the corresponding force, F, is calculated by the equation

$$F = \varepsilon\varepsilon_0 A/4 \cdot (U/h)^2, \tag{6.1}$$

where ε_0 is the vacuum permittivity ($\varepsilon_0 = 8.85 \cdot 10^{-12}$ F/m), ε is the relative permittivity (for air $\varepsilon \approx 1$), h is the levitation height. For electrodes numbered "1" and "3", the areas are $8.0 \cdot 10^{-7}$ and $4.3 \cdot 10^{-7}$ m^2, respectively. Results of measurements using electrodes "3" are shown in Fig. 6.3b. These two measurements help us to calculate the vertical and angular components of the actuator stiffness related to its centre using the approach given in Sect. 5.1.2. The results are presented in Table 6.1, measurement I, together with results of modelling and experimental parameters. Note that for modelling the analytical model of the inductive actuator developed in Sect. 5.1.3 is used. Also, the angular positioning of the disc can be performed in a range of ± 2 degrees. Obviously, if the levitation height is decreased, the range of the angular positioning is also decreased proportionally and vice versa.

As seen from the operation principle of the actuator, the positioning in vertical direction can be performed by changing the coils current and voltage applied to the suspending electrodes "1". Using the electrostatic force generated by electrodes "1", the disc can be held at a desired position while increasing the current value. As a result, the angular component of stiffness for the particular disc diameter is increased as predicted by the analytical model discussed above in Sect. 5.1.3. In order to demonstrate an increase of the angular component of stiffness, the coils were fed by the rms current value of 115 mA and applying the voltage of 42 V to the suspending electrodes "1", the disc was held at a levitation height of 100 mm. The results of measurements are shown in Table 6.1, measurement II. The analysis of Table 6.1 indicates that the angular component of the stiffness is increased two times compared with the same levitation height without electrostatic suspension (see Table 6.1, measurement III). Now we repeat the experiment with a disc having a diameter of 3.2 mm. Results of measurements and modelling are shown in Table 6.2. It is also worth noting that the experimental and modelling results are in good agreement.

Table 6.1 Results of measurements and modelling for disc diameter of 2.8 mm

Measurement	I	II	III
Measured Stiffness:			
Angular, [N · m·rad^{-1}]	0.6×10^{-8}	2.1×10^{-8}	1.0×10^{-8}
Vertical, [N · m^{-1}]	2.6×10^{-2}	2.6×10^{-2}	2.4×10^{-2}
Modelling Stiffness*:			
Angular, [N · m · rad^{-1}]	0.4×10^{-8}	0.9×10^{-8}	0.5×10^{-8}
Vertical, [N · m^{-1}]	2.4×10^{-2}	2.4×10^{-2}	2.3×10^{-2}
Parameters:			
Levitation Height, [mm]	120	100	100
Voltage on Electrodes "1", [V]	0	42	0
Coils Current, [mA]	115	115	109

Table 6.2 Measurement and modelling for disc diameter of 3.2 mm

Measurement no.	I	II
Measured stiffness		
Angular, [Nm · rad^{-1}]	$1.1 \cdot 10^{-8}$	$0.7 \cdot 10^{-8}$
Vertical, [N · m^{-1}]	0.040	0.040
Modelled stiffness		
Angular, [Nm · rad^{-1}]	$0.8 \cdot 10^{-8}$	$0.5 \cdot 10^{-8}$
Vertical, [N · m^{-1}]	0.042	0.042
Parameters		
Levitation height, [μm]	100	70
Voltage on electrodes "1", [V]	0	54
Coil current, [mA]	115	115

In Fig. 6.4, we provide a contextual overview by consolidating the experimental results for a disc with 3.2 mm diameter, as well as the results for a lighter disc with 2.8 mm diameter, on the same map of stiffness components elaborated using our analytical model are provided. Integrating the electrode structure offers the possibility to modulate the vertical positioning of the disc not only by changing the current in the coils but also by applying a voltage to the electrodes; in this case, the electrodes are labelled "1". Using the electrostatic force generated by applying a voltage on electrodes "1", the disc can be maintained at a certain height even when the current in the levitation coil is increased, therefore changing the angular stiffness of the respective disc. This is demonstrated by applying a voltage of 54 V on electrodes "1" and recording a levitation height of 70 μm, as opposed to 100 μm for the same current in the coil structure, but no voltage applied to the electrodes.

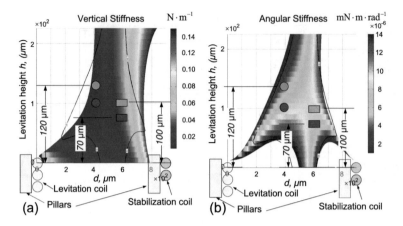

Fig. 6.4 Mapping of the distribution of stiffness components within the stability domain (according to the model developed in Sect. 5.1.3: **a** vertical stiffness; **b** angular stiffness. Circular/rectangular marks correspond to the proof mass with 2.8/3.2 mm diameter. Red/orange corresponds to the case when electrodes "1" are biased/not biased

The analytical model presented in Sect. 5.1.3 predicts a very slight change in the vertical stiffness when electrodes "1" are biased and the current in the coils is maintained constant. This is confirmed by the experimental results and reported in Tables 6.2 and 6.1: the vertical stiffness components both with and without applying the additional electrostatic force are $0.040 \, \text{N} \cdot \text{m}^{-1}$ and $0.026 \, \text{N} \cdot \text{m}^{-1}$ for the 3.2 mm and 2.8 mm diameter disc, respectively. The analytical model also predicts that the variation of the angular stiffness depends on a diameter of the disc. For the 2.8 mm diameter disc, the angular stiffness increases upon biasing electrodes "1" ($0.9 \cdot 10^{-8}$ versus $0.4 \cdot 10^{-8} \text{Nm} \cdot \text{rad}^{-1}$), whereas for the 3.2 mm diameter proof mass, the angular stiffness decreases upon biasing electrodes "1" ($0.5 \cdot 10^{-8}$ versus $0.8 \cdot 10^{-8} \text{Nm} \cdot \text{rad}^{-1}$). This excellent agreement between the theoretical predictions and the experiment is synthetically presented in Fig. 6.4.

Employing electrodes "2" the two phases were realized, which provides the oscillation of the disc with angular amplitude up to 37 degrees with the frequency of 1.5 Hz and applied voltage of ± 10 V as shown in Fig. 6.5a. Another important operating principle of this device is that it can operate as a bistable or pull-in actuator. Note that below the detailed analysis of pull-in phenomena in the hybrid actuator will be conducted including its analytical and quasi-FEM modelling in Sects. 6.1.3 and 6.1.5 below.

Since, in addition to the state of stable levitation, the actuator has two trivial stable states with a mechanical contact to the silicon surface. One state corresponds to the angular position when the disc edge comes in contact with the silicon surface as shown in Fig. 6.5b, which can be used for the extension of a range for angular positioning. The second state corresponds to the contact of the surfaces between the disc and silicon. The latter state can be used to protect the actuator against a

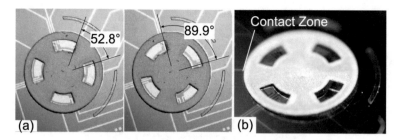

Fig. 6.5 a Angular oscillation with the amplitude up to 37 degrees and frequency 1.5 Hz at the levitation height to be 90 mm (this is a link on YouTube video: https://youtu.be/KzcJV2cSFrY); **b** Bistable state: the angular position when the PM edge comes in contact with the silicon surface

harsh environment like overloading, to keep the disc on the origin position. In order to drive the pull-in actuation, a voltage larger than the value corresponding to the pull-in effect is applied to the appropriate electrodes. Hence, energizing an electrode "3" and an electrode "1", the bistable actuation with angular positioning as depicted in Fig. 6.5b is performed. Because the number of electrodes "3" is four, hence, we have four directions for such the actuation. In a similar way, energizing electrodes "1" the bistable actuation along vertical direction is driven.

6.1.3 Eddy Current Simulation

In this section, a distribution of induced eddy current within a disc having a diameter of 2.8 mm levitated at height, h_l, of 200 μm by the coil structure corresponding to the design of hybrid actuator described above in the previous section is simulated by the developed quasi-FEM. According to the proposed procedure, the disc is homogenously meshed by 3993 circular elements. A result of meshing is a map of the location of elements with respect to the coordinate frame $\{x_k\}$ ($k = 1, 2, 3$), the origin of which is placed at the centre of the disc, as shown in Fig. 6.6. Each element crosses its neighbouring element only at one point. Depending on the location of neighbouring element at the top, right, bottom and left sides, this point can be placed on the element perimeter at an angle of $0°, 90°, 180°$ and $270°$ subtended at the centre of the circular element, respectively. Elements are numbering from left to right in each line. Because of the plane shape of the levitated micro-object, vectors of the list of elements $\{^{(s)}\underline{C}\}$ have the following structures, namely, $^{(s)}\rho = [^{(s)}x_1 \; ^{(s)}x_2 \; 0]^T$ and $^{(s)}\phi = [0 \; 0 \; 0]^T$ ($s = 1, \ldots, n$). Knowing $\{^{(s)}\underline{C}\}$, the $\underline{L}$ matrix can be calculated by Eq. (4.6).

3D geometry of two micro-coils is approximated by a series of circular filaments. Hence, depending on the number of windings, the levitation coil is replaced by 20 circular filaments, while the stabilization coil by 12 circular filaments. Thus, the total number of circular filaments, N, is 32. Assigning the origin of the fixed frame

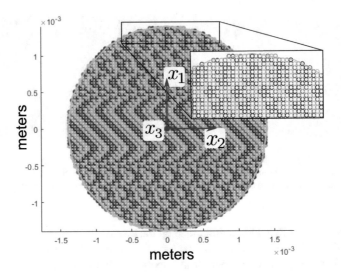

Fig. 6.6 Disc of a diameter of 2.8 mm is meshed by 3993 circular elements

$\{X_k\}$ ($k = 1, 2, 3$) to the centre of the circular filament corresponding to the first top winding of the levitation coil, the linear position of the circular filaments of levitation coil can be defined as $^{(j)}\underline{r}_c = [0\ 0\ (j-1) \cdot p]^T$, ($j = 1, \ldots, 20$), where p is the pitch equaling to 25 μm. The same is applicable for stabilization coil, $^{(j)}\underline{r}_c = [0, 0, (j-21) \cdot p]^T$, with the difference that the index j is varied from 21 to 32. For both coils, the Bryan angle of each circular filament is $^{(j)}\underline{\phi}_c = [0\ 0\ 0]^T$, ($j = 1, \ldots, 32$). Accounting for the values of diameters of levitation and stabilization coils, 3D geometrical scheme of the micro-actuator for eddy current simulation can be build as shown in Fig. 6.7.

The position of the coordinate frame $\{x_k\}$ ($k = 1, 2, 3$) with respect to the fixed frame $\{X_k\}$ ($k = 1, 2, 3$) is defined by the radius vector $\underline{r}_{cm} = [0\ 0\ h_l]^T$. Then, the position of the s-mesh element with respect to the coordinate frame $\{^{(j)}z_k\}$ ($k = 1, 2, 3$) assigned to the j-coil filament can be found as $^{(s,j)}\underline{r} = \underline{r}_{cm} + {}^{(s)}\underline{\rho} - {}^{(j)}\underline{r}_c$ or in a matrix form as

$$^{(s,j)}\underline{r}^z = {}^{(j)}\underline{A}^{zX}\underline{r}^X_{cm} + {}^{(j)}\underline{A}^{zx\,(s)}\underline{\rho}^x - {}^{(j)}\underline{A}^{zX\,(j)}\underline{r}^X_c, \tag{6.2}$$

where $^{(j)}\underline{A}^{zX} = {}^{(j)}\underline{A}^{zX}\left({}^{(j)}\underline{\phi}_c\right) = {}^{(j)}\underline{e}^z \cdot \underline{e}^X$ and $^{(j)}\underline{A}^{zx} = {}^{(j)}\underline{A}^{zX}\left({}^{(j)}\underline{\phi}_c\right)\underline{A}^{Xx}(\underline{\varphi}) = {}^{(j)}\underline{e}^z \cdot \underline{e}^x$ are the direction cosine matrices. Because all angles are zero, hence, $^{(j)}\underline{A}^{zx} = {}^{(j)}\underline{A}^{zX} = \underline{E}$, where $\underline{E}$ is the (3×3) unit matrix. Since the coils are represented by the circular filaments and using the radius vector $^{(s,j)}\underline{r}$, the mutual inductance between the j-coil and s-meshed element can be calculated directly by the formula (4.25) presented in Chap. 4. It is convenient to present the result of calculation in the dimensionless form. For this reason, the dimensionless currents in the levitation coil and stabilization one are introduced by dividing currents on the ampli-

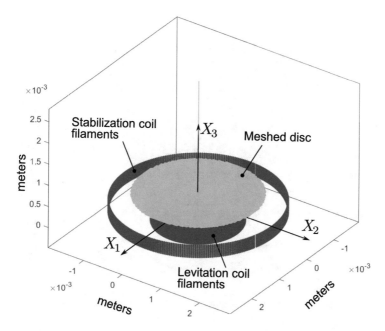

Fig. 6.7 3D geometrical scheme of the hybrid actuator for eddy current simulation: $\{X_k\}$ ($k = 1, 2, 3$) is the fixed coordinate frame

tude of the current in the levitation coil, since the amplitudes of the current in both coil are the same. Hence, the input current in the levitation coil filaments is to be one, while in the stabilization coil filaments to be minus one (because of the 180° phase shift). Now, the induced eddy current in dimensionless values can be calculated by Eq. (4.8).

To illustratively present the calculation result, the obtained (3993 × 1) eddy current matrix, $\underline{I}$, is transformed into the (71 × 71) 2D matrix, $\underline{I}$. Data in this (71 × 71) 2D matrix are allocated similarly to the structure corresponding to Fig. 6.6. Then, the distribution along the disc surface of induced eddy current in mesh circular elements are shown in Fig. 6.8. The analysis of Fig. 6.8 shows that in a central area of the disc, corresponding to the area of the circular cross-section of the levitation coil, the eddy current has the negative sign (it means that the direction of induced eddy current flow is opposite to the flow direction in the levitation coil) due to the significant contribution of the ac magnetic field generated by this coil, while outside of this area, a sign becomes positive due to the ac magnetic field of the stabilization coil.

Now let us present the obtained result in the vector form through unit vectors e_1^x and e_2^x of the base $\underline{e}^x$. Taking the numerical gradient of the (71 × 71) 2D matrix, $\underline{I}$ with respect to the rows and columns, the components in the form of the (71 × 71) 2D matrixes of $\underline{I}_1$ and $\underline{I}_2$ relative to the unit vectors e_1^x and e_2^x are calculated, respectively. Then, the (71 × 71) 2D matrix of magnitudes of the eddy current for each mesh point

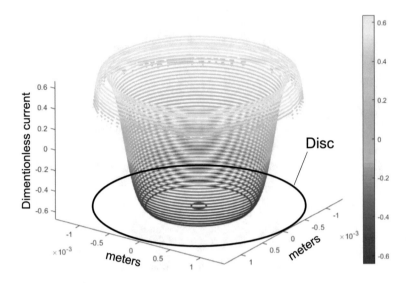

Fig. 6.8 The distribution of the eddy current in mesh circular elements

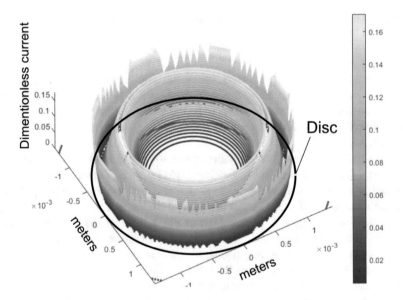

Fig. 6.9 The distribution of magnitudes of eddy current with respect to unit vectors of e_1^x and e_2^x of the base $\underline{e}^x$

is estimated by $\sqrt{\underline{I}_1{}^2 + \underline{I}_2{}^2}$. The result of the estimation is shown in Fig. 6.9. Figure 6.9 depicts that maximum magnitudes of eddy current are concentrated along the

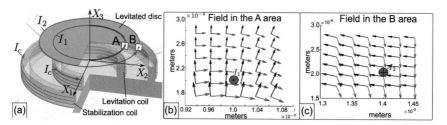

Fig. 6.10 Modelling of linear pull-in actuation: **a** scheme for modelling based on the approximation of induced eddy current in the disc by two eddy current circuits with I_1 and I_2 currents corresponding to its maximum magnitudes; **b** the map of magnetic field (blue arrows) and its corresponded gradient (black arrows) within the A area build on the X_3-X_2 plane around the I_1 circuit (see Fig. 6.10a); **c** the map of magnetic field and its corresponded gradient within the B area build on the X_3-X_2 plane around the I_2 circuit (see Fig. 6.10a)

edge of the disc and in its central part along the circle having the same diameter as the levitation coil. This result is similar to one obtained by Lu in work [2], where the induced eddy current in a disc levitated by two coils with the similar design was simulated by COMSOL software. Both results provide the reason and applicability of the approximation based on two eddy current circuits, for the analysis of HL-micro-systems with an axially symmetric design.

6.1.4 Analytical Model of Static Linear Pull-In Actuation

The approximation of induced eddy current within the disc based on two eddy current circuits is applied to model the linear pull-in actuation in hybrid levitation micro-actuator with the design under consideration. Scheme for modelling is shown in Fig. 6.10a. Due to particularities of the hybrid levitation micro-actuator design, a radius of the I_1 circuit is equal to 1 mm, while a radius of the I_2 circuit is restricted by the radius of disc equaling to 1.4 mm. Let us examine the field distribution around these two circles of eddy current circuits. Since the design is axially symmetric, it is enough to consider some areas such as denoted by A and B as shown in Fig. 6.10a cutting, for instance, on the X_3-X_2 plane and crossing the I_1 and I_2 circuit, respectively. Mapping the field in the A and B area shows that the force acting on the I_1 circuit determined by the corresponded gradient of the field is directed vertically and lifts up the disc, while the force acting on the I_2 circuit has almost horizontal direction and pushes the disc towards the centre as presented in Fig. 6.10b and c, respectively. Noting that the names of coils as levitation and stabilization coil reflect their functionalities coming to form the field analysis. It can be assumed that for a diameter of disc equalling to 2.8 mm the influence of the I_2 circuit on the linear pull-in actuation is small and can be neglected. Thus, modelling of the linear pull-in actuation is reduced to the force interaction between the currents in the levitation coil and the I_1 circuit.

Then, the behaviour of the disc along the X_3 axis can be described as follows:

$$m\frac{d^2q_3}{dt^2} + mg + \frac{I_c^2}{L^o}\frac{dM}{dq_3}M + \frac{A_0}{4}\frac{U^2}{(h+q_3)^2} = 0, \tag{6.3}$$

where U is the applied voltage to the electrodes (namely, electrodes 1 and 2 shown in Fig. 6.24b), which have the same area of A_e, $A_0 = \varepsilon_0 A_e$, ε_0 is the permeability of free space, h is the space between the electrode surface and the origin of the coordinate frame $\{x_k\}$ ($k = 1, 2, 3$) measured along the X_3 axis. For this particular case, the mutual inductance, M, can be defined by the Maxwell formula such as

$$k^2 = \frac{4R_l^2}{4R_l^2 + (h_l + y)^2};$$
$$M = \mu_0 R_l \left[\left(\frac{2}{k} - k \right) K(k) - \frac{2}{k} E(k) \right], \tag{6.4}$$

where μ_0 is the magnetic permeability of free space, R_l is the radius of levitation coil, K and E are complete elliptic integrals of the first and second kinds, respectively.

For further analysis, model (6.3) is presented in the dimensionless form as follows:

$$\frac{d^2\lambda}{d\tau^2} + 1 - \eta \left[\left(\frac{2}{k} - k \right) K(k) - \frac{2}{k} E(k) \right] \frac{2}{k^2}$$
$$\times \left[\frac{2-k^2}{2(1-k^2)} E(k) - K(k) \right] \cdot \frac{\kappa \xi^2 (1 + \kappa\lambda)}{(1 + \xi^2 (1 + \kappa\lambda)^2)^{3/2}} \tag{6.5}$$
$$+ \frac{\beta}{(1+\lambda)^2} = 0,$$

where $\tau = \sqrt{g/ht}$, $\lambda = q_3/h$, $\eta = I^2 a^2/(mghL^o)$, $\beta = A_0 U^2/(4mgh^2)$, $\kappa = h/h_l$, $a = R_l\mu_0$ and $\xi = h_l/(2R_l)$.

From Eq. (6.5), the static pull-in model is

$$\beta = (1 + \lambda)^2 \left(-1 + \eta \left[\left(\frac{2}{k} - k \right) K(k) - \frac{2}{k} E(k) \right] \frac{2}{k^2} \right.$$
$$\left. \times \left[\frac{2-k^2}{2(1-k^2)} E(k) - K(k) \right] \cdot \frac{\kappa \xi^2 (1 + \kappa\lambda)}{(1 + \xi^2 (1 + \kappa\lambda)^2)^{3/2}} \right), \tag{6.6}$$

where the η constant can be defined at equilibrium state of the system, when λ and β are zero, as follows:

$$\eta = \left(\left[\left(\frac{2}{k_0} - k_0 \right) K(k_0) - \frac{2}{k_0} E(k_0) \right] \frac{2}{k_0^2} \right.$$
$$\left. \times \left[\frac{2-k_0^2}{2(1-k_0^2)} E(k_0) - K(k_0) \right] \cdot \frac{\kappa\xi^2}{(1+\xi^2)^{3/2}} \right)^{-1}, \tag{6.7}$$

and $k_0^2 = 1/(1 + \xi^2)$.

6.1.5 Quasi-FEM of Static Linear Pull-In Actuation

The quasi-FEM model of the static linear pull-in has a similar form to (6.3). The difference arises due to the fact that the magnetic interaction between the disc and coils along the X_3 axis is defined by the force F_3 from Eq. (4.9). Hence, taking into account this fact, the quasi-FEM model becomes

$$m\frac{d^2 q_3}{dt^2} + mg + \underline{I}^T \frac{\partial M_c}{\partial q_3} \underline{I}_c + \frac{A_0}{4}\frac{U^2}{(h+q_3)^2} = 0. \tag{6.8}$$

Now, we present Eq. (6.8) in dimensionless from:

$$\frac{d^2\lambda}{d\tau^2} + 1 + \eta_0 F_m(\lambda) + \frac{\beta}{(1+\lambda)^2} = 0;$$

$$F_m(\lambda) = \sum_{s=1}^{n}\sum_{j=1}^{N} \eta_{sj}\frac{\partial \overline{M}_{sj}(\overline{x}_1, \overline{x}_2, (1+\lambda\kappa)\chi)}{\partial\lambda}, \tag{6.9}$$

where $\eta_0 = \mu_0 I_{c1}^2\sqrt{R_{c1}R_e}/(mgR_e)$, R_{c1} is the radius of the first winding of the levitation coil, $\eta_{sj} = \overline{I}_s\overline{I}_{cj}\sqrt{\overline{R}_{cj}}/\chi$, $\overline{I}_s = I_s/I_{c1}$ and $\overline{I}_{cj} = I_{cj}/I_{c1}$ are the dimensionless currents, $\overline{R}_{cj} = R_{cj}/R_{c1}$, $\chi = h_l/R_e$ is the scaling factor, $\partial\overline{M}_{sj}/\partial\lambda$ is the derivative of dimensionless mutual inductance with respect to λ (its definition is shown below), $\overline{x}_1 = x_1/R_e$ and $\overline{x}_2 = x_2/R_e$ are the dimensionless coordinates. Noting that x_1 and x_2 are defined by Eq. (6.2).

The static pull-in model based on quasi-FEM (6.9) is

$$\beta = -(1+\lambda)^2\left(1 + \eta_0 F_m(\lambda)\right), \tag{6.10}$$

where similar to (6.7) the η_0 constant is also defined at equilibrium state as follows:

$$\eta_0 = -1/F_m(0). \tag{6.11}$$

The Derivative of Dimensionless Mutual Inductance with Respect to λ

Due to the particularity of the problem, namely, there is no angular misalignment between a circular element and a coil. Hence, the original formula for calculation of the mutual inductance between two circular filaments based on the Kalantarov-Zeitlin approach [3] in a dimensionless form can be rewritten as follows:

$$\overline{M}_{sj} = \frac{1}{\pi}\int_0^{2\pi}\frac{1+\overline{x}_1\cdot\cos\varphi + \overline{x}_2\cdot\sin\varphi}{\overline{\rho}^{1.5}}\frac{\Psi(k)}{k}d\varphi, \tag{6.12}$$

where

$$\overline{\rho} = \sqrt{1 + 2(\overline{x}_1\cdot\cos\varphi + \overline{x}_2\cdot\sin\varphi) + \overline{x}_1^2 + \overline{x}_2^2}; \tag{6.13}$$

$$\Psi(k) = \left(1 - \frac{k^2}{2}\right) K(k) - E(k); \tag{6.14}$$

$$k^2 = \frac{4\nu_j \bar{\rho}}{(\nu_j \bar{\rho} + 1)^2 + \nu_j^2 \bar{x}_3^2}, \tag{6.15}$$

where $\nu_j = R_e/R_{cj}$, R_{cj} is the radius of the j-coil filament, $\bar{x}_1$, $\bar{x}_2$ and $\bar{x}_3$ are the components of the radius vector $\boldsymbol{r}$ in base $\underline{e}^z$ (see Eq. (6.2)).

The derivative of dimensionless mutual inductance with respect to $\bar{x}_3$ is

$$\frac{\partial \overline{M}_{sj}}{\partial \bar{x}_3} = \frac{1}{\pi} \int_0^{2\pi} \frac{1 + \bar{x}_1 \cdot \cos \varphi + \bar{x}_2 \cdot \sin \varphi}{\bar{\rho}^{1.5}} \Phi(k) d\varphi, \tag{6.16}$$

where

$$\Phi(k) = \frac{d}{d\bar{x}_3} \frac{\Psi(k)}{k} = \frac{1}{k^2} \left(\frac{2 - k^2}{2(1 - k^2)} E(k) - K(k)\right) \frac{dk}{d\bar{x}_3}, \tag{6.17}$$

$$\frac{dk}{d\bar{x}_3} = -\frac{\nu_j^2 \bar{x}_3 \sqrt{4\nu_j \bar{\rho}}}{\left((1 + \nu_j \bar{\rho})^2 + \nu_j^2 \bar{x}_3^2\right)^{3/2}}. \tag{6.18}$$

Substituting $\bar{x}_3 = \lambda \kappa \chi$ into Eq. (6.16), the desired equation for the derivative of dimensionless mutual inductance with respect to λ is derived.

6.1.6 Preliminary Analysis of Developed Models

If the magnetic field gradient in the B area around the I_2 circuit of eddy current within the disc and corresponding force is directed almost horizontally (see Fig. 6.10c), then the estimation of pull-in parameters by means of the analytical model (6.6) becomes close to the exact calculation performed by the quasi-FEM (6.10). This fact indicates that the application of the analytical model requires the knowledge about the gradient of the magnetic field in the B area of a particular design under consideration. On the other hand, due to design particularities of the hybrid actuator, there are some particular cases, which can be immediately treated by the model (6.6) presenting a solution in simple analytical equations. In turn, these simple equations are convenient for practical application.

Indeed, upon holding a certain condition, for instance, if the design dimensionless parameter ξ is small, then the Maxwell formula can be well approximated by the logarithmical function. Hence, model (6.5) can be rewritten as follows:

$$\frac{d^2\lambda}{d\tau^2} + 1 - \eta \frac{\kappa}{1 + \kappa\lambda} \left[\ln \frac{4}{\xi(1 + \kappa\lambda)} - 2\right] + \frac{\beta}{(1 + \lambda)^2} = 0,. \tag{6.19}$$

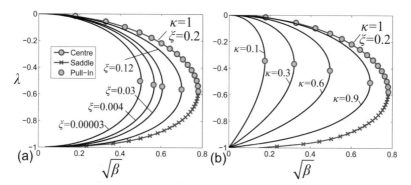

Fig. 6.11 Bifurcation diagrams and their evolution in depending on the design parameters: **a** the effect of ξ changing in a range from 3×10^{-5} to 0.2; **b** the effect of κ in a range from 0.1 to 1.0 (centre and saddle are corresponding to stable and unstable equilibrium, respectively)

As it is shown in Sect. 6.3 below, the accuracy of approximation is dependent on the parameter ξ. If the parameter ξ is less than 0.3, the electromagnetic force is approximated by the logarithmic function with the error less than six percentages. Upon trending the parameter ξ to the zero, the error between the exact equation and approximation as well trends to the zero. Worth noting that in all known prototypes of HL-micro-systems published in the literature the parameter ξ is less than 0.25. This fact provides the applicability of the reduced model for further analytical study. At the equilibrium point ($\lambda = \beta = 0$), η must be equal to $D = \ln 4/\xi - 2$. Hence, the equilibrium state is defined by the following equation:

$$f(\lambda, \beta) \equiv -\frac{\kappa\lambda}{1 + \kappa\lambda} - \frac{\ln(1 + \kappa\lambda)}{D(1 + \kappa\lambda)} - \frac{\beta}{(1 + \lambda)^2} = 0. \tag{6.20}$$

Using (6.20), the bifurcation diagram can be mapped as shown in Fig. 6.11, which depicts the distribution of centre and saddle as well as pull-in points depending on the design parameters. The pull-in points correspond to the transient state, in which the sign of $f(\lambda, \beta)$ is unchangeable in the vertical direction [4]. For a case of $\kappa = 1$, pull-in has the following parameters: displacement is $\lambda_{pi} = (1 - e - D)/(2D + e)$ (e is the Euler number), the square of voltage is $\beta_{pi} = -(\lambda_{pi} + \ln(1 + \lambda_{pi})/D)(1 + \lambda_{pi})$. For a case, when κ is small ($\kappa \ll 1$), pull-in parameters can be approximated as $\lambda_{pi} \approx -1/(3 - 2\kappa)$ and $\beta_{pi} \approx \kappa(1 + 1/D)(2 - 2\kappa)^2/(3 - 2\kappa)^3$.

Once κ tends to zero, it leads to the possibility of the further linearization of the magnetic force in the analytical model (6.6). From a physical point of view, it means that the redistribution between energy stored within the electrical field of capacitors and energy stored within magnetic field of coils and levitated disc occurs by changing the location of electrodes along the X_3 axis and, in particular, the electrodes are located closer to the levitated disc. Thus, we can write the following simple model of static pull-in actuation such as

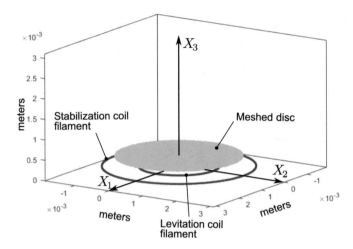

Fig. 6.12 The 3D scheme for simulation: the 3.1 mm diameter disc is meshed by 3993 circular elements

$$\beta = -\frac{\ln(4/\xi) - 1}{\ln(4/\xi) - 2}\kappa\lambda(1 + \lambda)^2. \tag{6.21}$$

From the latter model, the pull-in parameters can be estimated to be

$$\lambda_p = -\frac{1}{3}, \quad \beta_p = \frac{\ln(4/\xi) - 1}{\ln(4/\xi) - 2}\kappa\frac{4}{27}. \tag{6.22}$$

From the analysis of (6.22), we see that the pull-in displacement becomes the same as in classic static pull-in occurring in the spring-mass system with electrostatic actuation [5]. However, the square of pull-in voltage is different from the classical one and becomes smaller than the ratio of 4/27.

Now let us apply the obtained models, namely, (6.6), (6.10) and (6.21) to the design shown in Fig. 6.12, which consists of two circular plane coils having radii of 1.0 mm and 1.9 mm for levitation and stabilization coil, respectively. The disc is levitated at height, h_l, of 250 μm, while the electrodes are placed at a point measured from the equilibrium point O of the disc along the X_3 axis on a distance, h, of 10 μm. Hence, the dimensionless parameters of the design become $\kappa = 0.04$ and $\xi = 0.125$. Then, according to Eq. (6.22), the square dimensionless pull-in voltage can be calculated and becomes $\beta_p = 1.4930\frac{4}{27} = 0.01$. The modelling is performed for a disc with the following radii: 1.2, 1.55 and 1.7 μm. Figure 6.12 depicts the 3D scheme for simulation. The disc is meshed by 3993 circular elements.

The three equilibrium curves of pull-in actuation for a disc having a diameter of 1.55 μm calculated by models (6.6), (6.10) and (6.21) are shown in Fig. 6.13a. Results of estimation of pull-in parameters are summed up in Table 6.3. The simulation shows that the pull-in displacement (in absolute value) is $\lambda_p = 0.34$ and the pull-in voltage

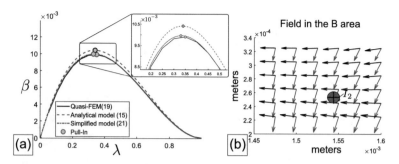

Fig. 6.13 Pull-in actuation of the 3.1 mm diameter disc: **a** equilibrium curve of the square voltage versus displacement (absolute value of λ); **b** the map of the magnetic field (blue arrows) and its corresponded gradient (black arrows) within the B area build on the X_3-X_2 plane around the I_2 circuit

Table 6.3 Pull-in parameters for the 3.1 mm diameter disc

Model	Pull-in parameter			Relative Error	
	λ_p	β_p	$\sqrt{\beta_p}$	$\Delta\lambda_p$	$\Delta\sqrt{\beta_p}$
Quasi-FEM, (6.10)	0.34	0.0099	0.0995	–	–
Analytical model, (6.6)	0.34	0.0104	0.102	0	0.025
Simplified model, (6.21)	0.3333	0.01	0.1	0.0474	0.005

is $\sqrt{\beta_p} = 0.0995$. The relative errors of estimation of pull-in parameters by means of the analytical models do not exceed 5% (please see Table 6.3). Noting that the map of gradient of the magnetic field in the B area is directed almost horizontally as shown in Fig. 6.13b. Hence, the contribution of electromagnetic force due to interaction between the magnetic field and the I_2 circuit to the pull-in actuation is small.

Figure 6.14a shows the equilibrium curve of pull-in actuation for a disc having a diameter of 2.4 mm, which is simulated by quasi-FEM (6.10). Since the analytical models (6.6) and (6.21) are independent of a radius of the levitated disc, the results of modelling are the same as presented in Fig. 6.13a and Table 6.3. The simulation predicts the following values of the pull-in parameters such as

$$\lambda_p = 0.34, \quad \beta_p = 0.01653, \quad \sqrt{\beta_p} = 0.1286, \tag{6.23}$$

where the dimensionless displacement λ_p is given in absolute value. Although the pull-in displacement has the same value as in the previous example, the main difference appears in the estimation of the pull-in voltage, which is increased due to the contribution of the electromagnetic force exerted on the I_2 circuit as shown in Fig. 6.14b. As a result, the relative error of the calculation of pull-in voltage by means of the analytical models is also increased drastically to 30%.

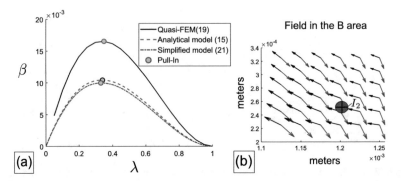

Fig. 6.14 Pull-in actuation of the 2.4 mm diameter disc: **a** equilibrium curve of the square voltage versus displacement (absolute value of λ); **b** the map of the magnetic field (blue arrows) and its corresponded gradient (black arrows) within the B area build on the X_3-X_2 plane around the I_2 circuit

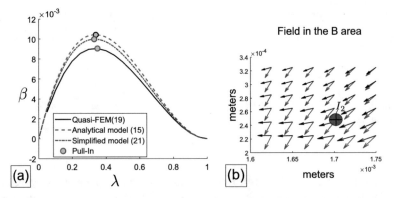

Fig. 6.15 Pull-in actuation of the 3.2 mm diameter disc: **a** equilibrium curve of the square voltage versus displacement (absolute value of λ); **b** the map of the magnetic field (blue arrows) and its corresponded gradient (black arrows) within the B area build on the X_3-X_2 plane around the I_2 circuit

In the case of the 3.4 mm diameter disc, the equilibrium curve simulated by quasi-FEM (6.10) is shown in Fig. 6.15a. The simulation predicts the following pull-in parameters such as

$$\lambda_p = 0.34, \quad \beta_p = 0.009, \quad \sqrt{\beta_p} = 0.0949. \tag{6.24}$$

In this case, the calculation of pull-in voltage by means of analytical models shows that the relative error is around 5%. As seen, the value of the pull-in voltage is decreased due to the electromagnetic force acting on the I_2 circuit, which has a vertical component directed against the levitation force as shown in Fig. 6.15b. Noting that in all three cases considered above, the map of the magnetic field and the corresponded gradient in the A area is similar to the map shown in Fig. 6.10b.

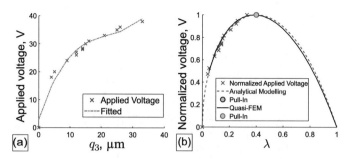

Fig. 6.16 Pull-in actuation of disc having a 2.4 mm diameter: **a** experimental measurement of applied voltage versus linear displacement; **b** normalized voltage versus dimensionless displacement: measurement data together with equilibrium curves generated by quasi-FEM and analytical one

6.1.7 Comparison with Experiment

Now let us compare the results of simulation and modelling generated by the quasi-FEM and analytical model, respectively, with experimental data collected for linear pull-in actuation performed by three discs having diameters of 2.4, 2.8 and 3.2 mm.

A Light Disc of A 2.4 mm Diameter

Figure 6.16a shows the result of the measurement of applied voltage to the electrodes 1 and 2 (see, Fig. 6.24b) against the displacement of the disc having a 2.4 mm diameter. The disc was levitated at a height of 100 μm metering from the electrode plane and its displacement was measured by the laser sensor. The pull-in actuation occurred at a height of 65 μm corresponding to the pull-in displacement of 35 μm upon applying 38 V to the electrodes. Figure 6.16b shows the comparison of equilibrium curves generated by quasi-FEM (6.10) and analytical model (6.6) with experimental measurement in normalized voltage. For the simulation, the 3D geometrical scheme as shown in Fig. 6.7 with the same dimensions for coils are used. The disc is meshed by 3993 elements as shown in Fig. 6.6. The modelling is carried out with the following dimensionless parameters: $\xi = 0.09$ and $\kappa = 0.55$. The analysis of Fig. 6.16b depicts a good agreement between both developed models itself. In particular, both models predict the same value of the pull-in displacement. Only, we see a slight difference between the shapes of the curves after the pull-in point. Also, both models in terms of normalized values are in good agreement with experiment. However, the comparison in terms of dimension values shown in Fig. 6.17 reveals that the analytical model gives a relative error, which is around 16% in the estimation of the pull-in voltage.

Disc of A 2.8 mm Diameter

Figure 6.18a presents the result of the measurement of pull-in actuation of the disc having a 2.8 mm diameter, which has been discussed above. The disc was levitated at a height of 120 μm measuring from the electrode plane. The pull-in actuation occurred at a height of 77 μm corresponding to the pull-in displacement of 43 μm

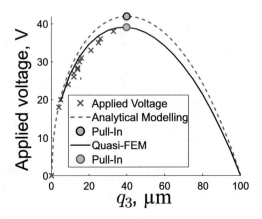

Fig. 6.17 Two equilibrium curves calculated by quasi-FEM Eq. (6.8) and analytical model Eq. (6.55) of applied voltage versus displacement of the disc having a 2.4 mm diameter in comparing with experimental data

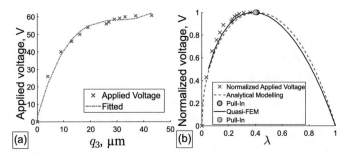

Fig. 6.18 Pull-in actuation of disc having a 2.8 mm diameter: **a** experimental measurement of applied voltage versus linear displacement (see, Sect. 6.1.2); **b** normalized voltage versus dimensionless displacement: measurement data together with equilibrium curves generated by quasi-FEM and analytical one

upon applying 60.8 V to the electrodes. The simulation is performed in a similar way as it was discussed in the previous Sect. 6.1.7. The difference is only in a size of diameter of the disc and a levitation height. For modelling, the following dimensionless parameters, namely, $\xi = 0.1$ and $\kappa = 0.6$ are used. Comparison of both models in normalized values as shown in Fig. 6.18b depicts a difference between the pull-in displacements predicted by quasi-FEM (6.10) and analytical model (6.6). The analytical model (6.6) gives a relative error, which is around 2% in estimation of the pull-in displacement. Also, there is a slight difference between the shapes of the curves after the pull-in point, similar to the previous result. Conducting the analysis in terms of dimension values as shown in Fig. 6.19, the relative error of around 13% produced by the analytical model in the calculation of the pull-in voltage can be recognized.

Fig. 6.19 Two equilibrium curves calculated by quasi-FEM Eq. (6.8) and analytical model Eq. (6.55) of applied voltage versus displacement of the disc having a 2.8 mm diameter in comparing with experimental data

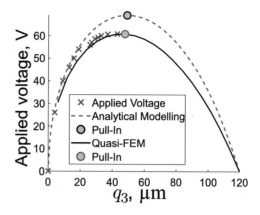

Disc of A 3.2 mm Diameter

The result of the experimental investigation of the pull-in actuation in the presented prototype of the hybrid levitation actuator with the disc of a 3.2 mm diameter was discussed and reported in our work [6]. Measuring from the electrode plane, the disc was levitated at a height of 107 μm. Figure 6.20a shows the measurement of applied voltage to the electrodes against the disc linear displacement along the vertical axis. The pull-in actuation occurred at a height of 71 μm corresponding to the pull-in displacement of 36 μm upon applying 65 V to the electrodes. The results of simulation and modelling are shown in Fig. 6.20b in the normalized values. The analysis of Fig. 6.20b reveals a slight shift on the right of the equilibrium curve generated by the analytical model (6.6) with respect to the curve modelled by the quasi-FEM (6.10).

Hence, this shift corresponds to the 13.5% relative error in the calculation of the pull-in displacement by the analytical model. The relative error produced by the analytical model in estimation of the pull-in voltage is calculated to be 30% as follows from the analysis of the results of simulation and modelling in dimension values shown in Fig. 6.21.

In addition, we add new data of measurement of the pull-in actuation performed by the same disc, which was levitated at a height of 64 μm. The result of the measurement of applied voltage against the linear displacement of the disc is shown in Fig. 6.22. The pull-in actuation occurred at a height of 46 μm corresponding the pull-in displacement of 18 μm upon applying the pull-in voltage equal to 32 V. Similar to Fig. 6.20b, the equilibrium curve generated by the analytical model has a slight shift on the right relative to the curve generated by the quasi-FEM as shown in Fig. 6.22b. This shift corresponds to the 9% relative error given by the analytical model in the estimation of the pull-in displacement. Comparing in terms of dimension values as shown in Fig. 6.23, the relative error of around 28% produced by the analytical model in the calculation of the pull-in voltage is appeared.

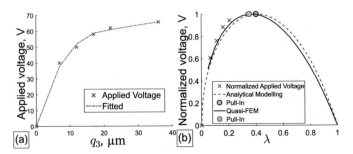

Fig. 6.20 Pull-in actuation of disc having a 3.2 mm diameter and levitated at 107 μm: **a** experimental measurement of applied voltage versus linear displacement [6]; **b** normalized voltage versus dimensionless displacement: measurement data together with equilibrium curves generated by quasi-FEM and analytical one

Fig. 6.21 Two equilibrium curves calculated by quasi-FEM Eq. (6.8) and analytical model Eq. (6.55) of applied voltage versus displacement of the disc having a 3.2 mm diameter and levitated at 107 μm in comparing with experimental data

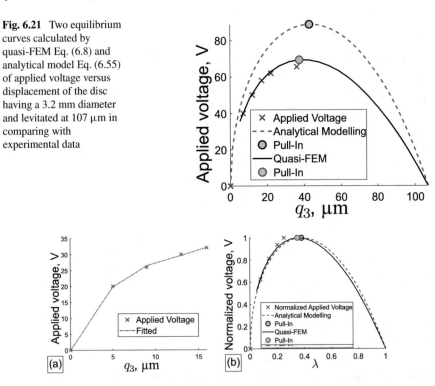

Fig. 6.22 Pull-in actuation of disc having a 3.2 mm diameter and levitated at 64 μm: **a** experimental measurement of applied voltage versus linear displacement; **b** normalized voltage versus dimensionless displacement: measurement data together with equilibrium curves generated by quasi-FEM and analytical one

Fig. 6.23 Two equilibrium
curves calculated by
quasi-FEM Eq. (6.8) and
analytical model Eq. (6.55)
of applied voltage versus
displacement of the disc
having a 3.2 mm diameter
and levitated at 64 μm in
comparing with
experimental data

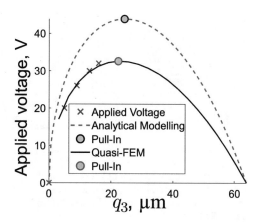

Table 6.4 Results of measurements and modelling of the static pull-in actuation

Measured parameters	Diameter of disc	2.4 mm	2.8 mm	3.2 mm	
	Mass of disc	0.2 mg	0.3 mg	0.7 mg	
	Levitation height, h_l	180 μm	200 μm	144 μm	187 μm
	Spacing, h	100 μm	119 μm	64 μm	107 μm
Calculated parameters	$\xi = h_l/(2R_l)$	0.09	0.1	0.072	0.0935
	$\kappa = h/h_l$	0.55	0.6	0.44	0.57
Measured pull-in parameters	**Displacement**	35 μm	43 μm	18 μm	36 μm
	Voltage	38 V	60.8 V	32 V	65 V
Pull-in parameters modelled by Eq. (6.55)	Displacement	40 μm	49 μm	24 μm	42 μm
	Voltage	43 V	69 V	44 V	88 μm
Pull-in parameters simulated by Eq. (6.8)	Displacement	40 μm	48 μm	22 μm	37 μm
	Voltage	37 V	69.8 V	33 V	69 V

All the results of measurements and modelling of the pull-parameters including
related particularities of the experiments and values of dimensionless parameters for
modelling are summed up in Table 6.4.

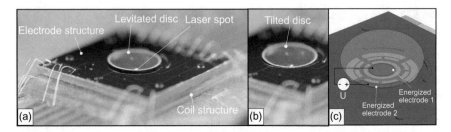

Fig. 6.24 Experimental test of tilting actuation in hybrid levitation micro-actuator and schemes for its modelling (this is a link on YouTube video: https://youtu.be/XfLMTru_vHI): **a** stable levitation of the disc having a diameter of 2.8 mm; **b** applying the pull-in voltage to the electrodes, the disc was tilted. The angle of tilting actuation is restricted by the mechanical contact with the surface of the electrode structure; **c** the set of electrodes (the energized electrodes are highlighted in red, U is the applied voltage)

6.1.8   Angular Pull-In

In this section, we focus on the study angular pull-in actuation performed in the hybrid levitation actuator prototype considered above. Similar to the linear pull-in, a disc with a diameter of 2.8 mm was levitated at different heights, however, the pull-in of tilting actuation was performed by applying pull-in voltage to electrodes 1 and 2 as shown in Fig. 6.24b, c. Using the same setup shown in Fig.6.2 and described in Sect. 6.1.2, the linear displacement of a disc edge was measured by a laser distance sensor directly (see, Fig. 6.24a, b).

Eddy Current

The position of the coordinate frame $\{x_k\}$ ($k = 1, 2, 3$) with respect to the fixed frame $\{X_k\}$ ($k = 1, 2, 3$) is defined by the radius vector $\underline{r}_{cm} = [0 \ 0 \ z]^T$, while its angular position is defined by the Bryan angles (Cardan angles) such as $\underline{\varphi} = [0 \ \theta \ 0]^T$. Hence, the variables z and θ are considered as generalized coordinates of the mechanical part of the actuator. Accounting for the values of diameters of levitation and stabilization coils, and locating the disc at a height of $z = h_l = 250 \, \mu m$ and tilting it on an angle of $10°$, the 3D geometrical scheme of the hybrid actuator for eddy current simulation can be built as shown in Fig. 6.25a.

The induced eddy current in each circular element is a solution of the following matrix equation (see, Chap. 4):

$$\underline{I} = \underline{L}^{-1} \underline{M}_c \underline{I}_c, \tag{6.25}$$

where $\underline{I}$ is the ($n \times 1$) matrix of eddy currents and $\underline{I}_c = [I_{c1} I_{c2} \dots I_{cN}]^T$ is the given ($N \times 1$) matrix of currents in coils. The matrix $\underline{L}$ can be formed as follows:

$$\underline{L} = L^o \underline{E} + \underline{M}^o, \tag{6.26}$$

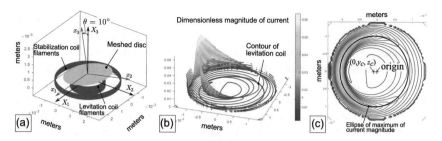

Fig. 6.25 Simulation of induced eddy current within the tilted disc: **a** 3D geometrical scheme of the hybrid actuator for eddy current simulation: the disc having a diameter of 2.8 mm is meshed by 3993 elements; **b** the distribution of the magnitude of eddy current within disc; **c** elliptic shape of the maximum magnitude of eddy current in a central part of the disc: y_c and z_c are the coordinates of the ellipse centre with respect to the coordinate frame $\{x_k\}$ ($k = 1, 2, 3$)

where $\underline{E}$ is the $(n \times n)$ unit matrix, $\underline{M}^o$ is the $(n \times n)$-symmetric hollow matrix whose elements are the mutual inductance between k- and s-finite circular elements, L_{ks}^o, ($k \neq s$) and, besides, $^{(k,s)}\rho = \,^{(k)}\rho - \,^{(s)}\rho$. The self-inductance of the circular element is calculated by the known formula for a circular ring with circular cross-section:

$$L^o = \mu_0 R_e \left[\ln 8/\varepsilon - 7/4 + \varepsilon^2/8 \, (\ln 8/\varepsilon + 1/3)\right], \qquad (6.27)$$

where μ_0 is the magnetic permeability of free space, $\varepsilon = th/(2R_e)$; th is the thickness of a mashed layer of micro-object and R_e is the radius of circular element. Knowing a projection of the j-coil filament loop on the plane of the k-circular element, the Kalantarov-Zeitlin method [3] can be used in order to calculate the mutual inductance. Hence, the $(n \times N)$ matrix $\underline{M}_c$ consisting of elements of the mutual inductance M_{kj} can be obtained.

It is convenient to present the result of the calculation in the dimensionless form. For this reason, the dimensionless currents in the levitation coil and stabilization one are introduced by dividing currents on the amplitude of the current in the levitation coil. Since the amplitudes of the current in both coils are the same. Hence, the input current in the levitation coil filaments is to be one ($I_{c1} = I_{c2} = \cdots = I_{c20} = 1$), while in the stabilization coil filaments to be minus one (because of the 180° phase shift) ($I_{c21} = I_{c2} = \cdots = I_{c32} = -1$). Now, the induced eddy current in dimensionless values can be calculated.

Then, the calculated (3993×1) eddy current matrix, $\underline{I}$, is transformed into the (71×71) 2D matrix, $\underline{I}$. Moreover, taking the numerical gradient of the obtained 2D matrix (71×71) of $\underline{I}$ with respect to the rows and columns, the components in the form of the (71×71) 2D matrixes of $\underline{I}_1$ and $\underline{I}_2$ relative to the unit vectors e_1^x and e_2^x are calculated, respectively. Consequently, the (71×71) 2D matrix of magnitudes of the eddy current for each mesh point is estimated by $\sqrt{\underline{I}_1^2 + \underline{I}_2^2}$. The result of estimation is shown in Fig. 6.25b.

Analysis of Fig. 6.25b depicts that two areas corresponding to the maximum magnitude of eddy current can be recognised. Similar to the disc without tilting [2], maximum magnitude of eddy current is concentrated along the edge of the disc and in its central part. However, from Fig. 6.25c, it is seen that, in this case, the maximum magnitude of eddy current in a central part of the disc is patterned as an ellipse with shifted centre. The shape of this ellipse and the location of its centre can be expressed as functions of the generalized coordinates, namely, z and θ, which can be defined from the scheme shown in Fig. 6.25. Application of these functions for calculation of mutual inductance between the levitation coil and the elliptic shape circuit of eddy current corresponding to the maximum magnitude and of the electromagnetic force and torque acting on the disc is shown below.

Model

Study of magnetic field generated by the coils and its corresponding gradient around the areas of the concentrations of induced eddy current in the design of the hybrid actuator under consideration conducted in Sect. 6.1.3 reveals that the arising magnetic force acting on the edge of the disc has almost horizontal direction and pushes it toward the centre of symmetry of coils. Hence, a vertical component of this force contributing to the tilting pull-in actuation is assumed to be small. Due to this fact the electromagnetic part of this system can be reduced to the interaction between the current in the levitation coil and the induced eddy current in a centre part of the disc having the elliptic shape circuit as shown in Fig. 6.25c. Since the diameter of disc is one order of magnitude larger than the space between electrodes and the disc, it can be further assumed that the energized two capacitors formed by electrodes 1 and 2 can be considered as plane ones connected in series. Therefore, these two capacitors can be approximated by the following equations:

$$C_1(z) = \frac{\varepsilon_0 A_1}{h + z} = \frac{C_{10}}{1 + z/h}, \quad C_2(z, \theta) = \frac{\varepsilon_0 A_2}{h + z + R_t \theta} = \frac{C_{20}}{1 + z/h + R_t \theta/h}, \tag{6.28}$$

where $C_{10} = \varepsilon_0 A_1/h$, $C_{20} = \varepsilon_0 A_2/h$, A_1 and A_2 are the areas of electrodes 1 and 2, respectively, ε_0 is the permeability of free space, h is the space between the electrode surface and the origin of the coordinate frame $\{x_k\}$ ($k = 1, 2, 3$) measured along the X_3 axis, R_t is the distance between the centre of the disc and applied electrostatic force generated by electrode 2. Thus, according to given assumptions, the following reduced analytical model of tilting actuation in the hybrid actuator can be proposed:

$$\begin{cases} m\dfrac{d^2 z}{dt^2} + mg + \dfrac{I_c^2}{L_d}\dfrac{dM}{dz}M + \dfrac{C_{10}U^2}{2(1+d)(1+z/h+d/(1+d)R_t\theta/h)^2} = 0, \\ J\dfrac{d^2\theta}{dt^2} + \dfrac{I_c^2}{L_d}\dfrac{dM}{d\theta}M + \dfrac{C_{10}U^2 d}{2(1+d)^2(1+z/h+d/(1+d)R_t\theta/h)^2} = 0, \end{cases} \tag{6.29}$$

where m is the mass of the disc, J is the moment of inertia of the disc about the axis lying on its equatorial plane, I_c is the current in the levitation coil, L_d is the effective self inductance of the disc, U is the applied voltage to the electrodes (namely, electrodes 1 and 2 shown in Fig. 6.24b), M is the mutual inductance between the circular coil and the elliptic shape circuit of the current corresponding to the maximum magnitude of induced eddy current within the disc and $d = A_1/A_2$ is the ratio of the area of electrode 1 to the area of electrode 2.

For further analysis, let us rewrite the set (6.29) in a dimensionless form as follows :

$$
\begin{cases}
\dfrac{d^2\lambda}{d\tau^2} + 1 + \eta\dfrac{d\overline{M}}{d\lambda}\overline{M} + \dfrac{\beta}{(1+\lambda+d/(1+d)\bar{\theta})^2} = 0, \\[4mm]
\dfrac{d^2\bar{\theta}}{d\tau^2} + p\eta\dfrac{d\overline{M}}{d\bar{\theta}}\overline{M} + p\dfrac{d}{1+d}\dfrac{\beta}{(1+\lambda+d/(1+d)\bar{\theta})^2} = 0,
\end{cases}
\tag{6.30}
$$

where $\tau = \sqrt{g/h}\,t$ is the dimensionless time, $\lambda = z/h$ is the linear dimensionless displacement, $\bar{\theta} = R_t\theta/h$ is the dimensionless angle, $\beta = C_{10}U^2/(2mgh(1+d))$ is the square dimensionless voltage, $\eta = I_c^2(R_l\mu_0)^2/(mghL_d)$ and $p = mR_t^2/J$ are the constants. The dimensionless mutual inductance M is defined as follows.

Mutual Inductance

Using the Kalantarov-Zeitlin method (see Chap. 4) and taking into account the geometrical scheme shown in Fig. 6.25c, the formula for calculation of the dimensionless mutual inductance between the circular coil and the circuit of the current corresponding to the maximum magnitude of induced eddy current within the disc can be defined as follows:

$$
\overline{M} = \frac{1}{\pi}\int_0^{2\pi} \frac{\bar{r} + t_1\cdot\cos\varphi + t_2\cdot\sin\varphi}{k\bar{\rho}^{1.5}}\cdot\bar{r}\cdot\Psi(k)d\varphi,
\tag{6.31}
$$

where

$$
\bar{r} = \frac{\cos\theta}{\sqrt{(\tan^2\theta + 1)\sin^2\varphi + \cos^2\theta\cos^2\varphi}};
$$

$t_1 = 0.5\bar{r}^2\tan^2\theta\sin(2\varphi)\cdot\bar{y}_c$; $t_2 = \bar{y}_c$; $\bar{\rho} = \sqrt{\bar{r}^2 + 2\bar{r}\cdot\bar{y}_c\sin\varphi + \bar{y}_c^2}$; $\bar{y}_c = -2\xi(1 + \kappa\lambda)\tan(\theta)/(\tan(\theta)^2 + 1)$, where $\xi = h_l/(2R_l)$ is the dimensionless geometrical parameter, and $\lambda = z/h$ is the dimensionless linear displacement; $\Psi(k) = \left(1 - \frac{k^2}{2}\right)K(k) - E(k)$. $K(k)$ and $E(k)$ are the complete elliptic functions of the first and second kind, respectively, and $k^2 = \bar{\rho}/((0.25\bar{\rho} + 1)^2 + \xi^2\bar{z}_c^2)$; while, the dimensionless coordinate $\bar{z}_c$ is defined as $\bar{z}_c = (1 + \kappa\lambda)/(\tan^2\theta + 1) + \bar{r}\sin\varphi\tan\theta/(2\xi)$. From Eq. (6.31), the derivatives of the mutual inductance with respect to dimensionless variables of λ and $\bar{\theta}$ are derived as well.

Static Pull-In Actuation

Accounting for set (6.30), a model describing the static pull-in actuation of tilting disc becomes

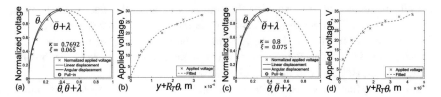

Fig. 6.26 Static pull-in of tilting actuation: **a** comparison of modelling with results of measurement I; **b** applied voltage versus the linear displacement of the disc for measurement I; **c** comparison of modelling with results of measurement II; **b** applied voltage versus the linear displacement of the disc for measurement II (other details are shown in Table 6.5)

$$
\begin{cases}
-\beta - \left(1 + \eta \dfrac{d\overline{M}}{d\lambda}\overline{M}\right)\left(1 + \lambda + d/(1+d)\bar{\theta}\right)^2 = 0, \\
-\beta - \eta \dfrac{d\overline{M}}{d\bar{\theta}}\overline{M}\dfrac{1+d}{d}(1 + \lambda + d/(1+d)\bar{\theta})^2 = 0.
\end{cases}
\tag{6.32}
$$

The η constant is defined from the initial state of the system, when β, λ and $\bar{\theta}$ are zero and must be equal to

$$
\eta = -1 \bigg/ \left(\dfrac{d\overline{M}}{d\lambda}\overline{M}\right).
\tag{6.33}
$$

Thus, the model (6.32) is the set of nonlinear equations determining the equilibrium state of the disc upon applying the voltage to the electrodes 1 and 2.

Model (6.32) is applied to mimic the experimental data measured in the setup described above. The results of measurements and modelling are shown in Fig. 6.26. Figure 6.26 depicts the comparison of modelling and experiment data in the normalized values. In particular, the total linear displacement estimated as a sum of linear and angular one measured during the experiment are presented in Figs. 6.26 and 6.27 reveals the direct comparison between measurements and modelling. The parameters of the device, particularities of conducted measurements and obtained results are summed up in Table 6.5. The analysis of Figs. 6.26, 6.27 and Table 6.5 shows that the modelling agrees well with experimental data.

6.2 Micro-Accelerators

The levitation micro-accelerating system based on the hybrid structure combining electric and inductive force field was proposed first time and demonstrated by Kraft's group [7]. Using the plane-symmetric design of coils as shown in Fig. 3.3a, Kraft's group propelled successfully a 7 μm thick micro-object of the 1×1 mm^2 size at the maximum levitation height of 82 μm and moved continuously the object at a maximum average velocity of 3.6 mm/s.

Table 6.5 Results of measurements and modelling of the static pull-in

	Parameters	Measurement I	Measurement II
Measured	Levitation height, h_l	130 μm	150 μm
	Spacing, h	100 μm	120 μm
	Pull-in displacement	**34 μm**	**45 μm**
	Pull-in voltage, U	**27 V**	**33 V**
Modelled	$\xi = h_l/2R_l$	0.065	0.075
	$\kappa = h/h_l$	0.7692	0.8
	Pull-in displacement	**38 μm**	**48 μm**
	Pull-in voltage, U	**28 V**	**33 V**
Device	Diameter of levitation coil, d_l	2mm	
	Area of electrodes, A_1 and A_2	0.8 and 0.43 mm^2	

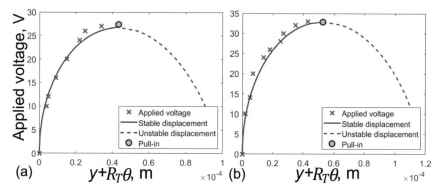

Fig. 6.27 Direct comparison of results of measurements and modelling (displacements are shown in absolute values): **a** for measurement I; **b** for measurement II (other details are shown in Table 6.5)

In this section, a micro-accelerator based on the same hybrid structure, however, in which instead of plane coils the 3D wire-bonded coils are used, is discussed. The design of the proposed device is shown in Fig. 6.28. The accelerator is also composed of two parts: a two-coil structure and a Si structure with electrodes. The coil structure consists of four straight tracks, and four semi-circles at the ends of the tracks. The long tracks are defined for the linear movement of a micro-object. The semi-circles are due to the nature of wire-bonding which requires a closed loop. The Si structure is with backside cavity and carries four electrode series on the front side, which are used to provide the electrostatic force. When the rectangular-shaped object is levitated, the electrodes are actuated with square waveforms to propel it.

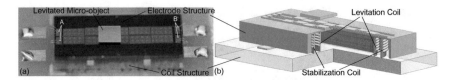

Fig. 6.28 Micro-accelerator (this is a link on YouTube video: https://youtu.be/jaAMS0B70OA): **a** the prototype of micro-accelerator glued onto a PCB under experimental testing: two electrode series for electrostatic actuation, the contact pads of which are denoted by letters A and B; **b** the scheme of the micro-transporter design

6.2.1 Operating Principle

During the operation of the device, two actuation mechanism coexist in the accelerator: the levitation force generated by the coils for the stable levitation of a micro-object, and the electrostatic propulsion force which moves it in one direction. The inductive levitation principle has been explained above chapters. Thus, only the electrostatic propelling mechanism is presented. The two pairs of contact pads (A and B) of electrode series shown in Fig. 6.28 are energized by a sequence of square impulses. The two electrode series with contact pairs A and B are symmetric in forward and backward directions. For propelling a micro-object, two sequence cycles of energizing of electrode series are performed. Only one pair (for instance, the A pair) is energized, and the other pair (the B pair) is with zero voltage. In the next half-cycle, the latter pair changes from zero voltage to be energized, and the former pair becomes deactivated.

The length of the rectangular-shaped object is the sum of the length of electrode and the gap between two adjacent electrodes. The electric field formed by two adjacent electrodes generates the force to propel it. However, the electrodes not only generate the propelling force. The electrode series attract also the micro-object. This attraction force is unwanted. Therefore, the voltage applied to the electrode can only be within a limited amplitude domain; otherwise, the vertical force might result in pull-in effect, by which the micro-object is adhered to one or more electrodes, causing potential problems such as short circuit or adhesion.

6.2.2 Micro-Fabrication

The fabrication of the accelerator consists of two separate parts similar to the fabrication process of hybrid micro-actuator described in Sect. 6.1, namely, micro-fabrications of the coil structure and of the electrode structure. The fabrication of the coil structure is of the same procedure as previously discussed in Chap. 2. To define the wire-bonding trajectory, a new Matlab code has been written for this purpose, which generates a winding trajectory. The new code splits each loop into four steps: two half circles and two long tracks. The duration and the speed of the capillary dur-

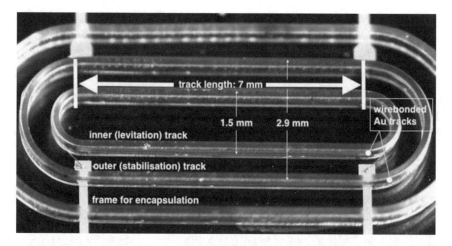

Fig. 6.29 Maglev tracks fabricated by automatic wire bonding of 25 μm diameter insulated Au wire, before encapsulation and before being integrated with the electrode structure. The inner track is 1.5 mm wide, has 14 windings and provides levitation (lift) of the conductive proof mass. The outer track is 2.9 mm wide, has ten windings and provides the stabilization of the conductive proof mass. The outer SU-8 frame is reserved for epoxy encapsulation [8]

ing each step were tuned to avoid fabrication failures. The wire-bonded coil structure is shown in Fig. 6.29. The track length is defined as 2.8 mm, 2.5 times of the length of the rectangular-shaped levitated micro-object. The diameters are 1.5 mm and 2.9 mm for the inner coil and the outer coil, respectively. Accounting for results of the analysis of the coil impedance and the levitation height as a function of frequency discussed in Chap. 5, the windings in the inner coil and the outer coil have been proposed to be 14 and 10, respectively. Thus, the device is able to provide enough levitation height of a micro-object, while remaining relatively low impedance around 6 MHz. During experiment, it has been observed that the wires along the tracks repel each other because of thermal stress, resulting in high tension in the coil, and wedge peel off. Therefore, an extra ball-wedge bond is added to protect the wedge bond in case of wedge peel off during experiment with the ball bond overlapping with the wedge bond of the coil. By adding the protect ball bond over the wedge bond of coil, device failure due to wedge peel off has been significantly reduced during the experiment. The fabrication of the Si structure is exactly the same as the fabrication process described for hybrid levitation micro-actuators in Sect. 6.1. Finally, two fabricated structures assembled into one device shown in Fig. 6.28a by a flip on-chip machine. Figure 6.28a shows the stable levitation of the square shape of object having the size of 2.4×2.4 mm^2.

Fig. 6.30 Captures from a movie showing transportation of a disc-shaped object along the axis of the track due to tilting. The coils have been encapsulated with a thermally conductive, electrically insulating epoxy (black) to efficiently evacuate the heat generated in the coils [8] (this is a link on YouTube video: https://youtu.be/Ii4TWRyWjfY)

6.2.3 Linear Motion Due to the Gravity

The fabricated device allows us to levitate stable different shapes of micro-objects in addition to rectangular shaped object, the levitation of disc-shaped object was experimentally demonstrated. In particular, Fig. 6.30 shows the movement of a disc-shaped object along the axis of the tracks due to device tilting without the electrode structure. The track axis is a locus of neutral equilibrium with respect to the direction parallel to the track axis and a locus of stable equilibrium with respect to the direction perpendicular to the track axis. The disc stops at the position where the Lorentz force and the weight reach equilibrium. Within a tiling angle of 20°, the disc can remain stably levitated. Once the angle is over 20°, the disc flips out from either end of the track axis. It has been found during the experimental procedure that, for the disc, two stable equilibrium positions exist at the ends of the tracks.

6.2.4 Modelling of Stable Levitation

In this section, the stability of the prototype operated in air environment for the case of a rectangular-shaped levitated micro-object is studied. According to the procedure proposed above, firstly, a representative eddy current circuit is defined. Then the equations for coefficients of Taylor series (3.39) are computed. The distribution of eddy currents generated by the micro-coils can be studied using a similar design of the prototype consisting of four straight wires and a rectangular-shaped micro-object. Taking into account that the levitation height of the micro-object is significantly smaller than its lateral dimensions, the eddy current distribution can be represented as shown in Fig. 6.31b. The simulation was performed for a levitation height of 100 μm and coil currents of 100 mA. The distribution in Fig. 6.31a is presented in dimensionless relative values, i.e., the ratio of the current density to its maximum value. The analysis of Fig. 6.31a shows that the representative eddy current circuit can consist of two circuits as shown in Fig. 6.31b covering a particular eddy current density range between 0.42 and 1.0. It is important to note that the behaviour of the eddy current circuit, i_2, is similar to the one in the axially symmetric design and its

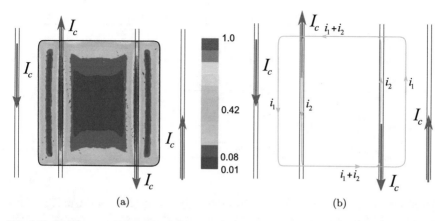

Fig. 6.31 Eddy current induced in the rectangular micro-object (Add results of own simulation): **a** distribution of eddy current density; **b** representative circuit

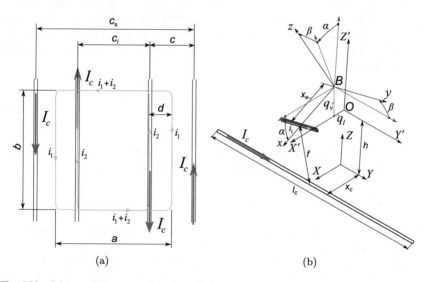

Fig. 6.32 Scheme of transporter for calculation

position in space does not depend on the lateral displacement of the micro-object characterized by the generalized coordinate, q_l.

The scheme for the calculation of the stiffness and stability of the transporter is shown in Fig. 6.32a. Using equations for self-inductances of a rectangle and straight wire having a square cross-section [9, pp. 320 and 315], respectively, and mutual inductance of two parallel wires [9, p. 306], terms L_{ks}^0 of (3.39) can be defined as follows:

$$L_{11}^0 = \frac{\mu_0}{\pi} \cdot (c_l + 2d + b) \left[\ln \frac{2(c_l + 2d)b}{\chi} - \frac{(c_l + 2d)\ln(c_l + 2d + b) + b\ln(c_l + 2d + b)}{c_l + 2d + b} \right.$$
$$\left. + \frac{\sqrt{(c_l + 2d)^2 + b^2}}{c_l + 2d + b} - \frac{1}{2} + 0.477 \frac{\chi}{c_l + 2d + b} \right];$$

$$L_{22}^0 = \frac{\mu_0}{\pi} \cdot (c_l + b) \left[\ln \frac{2c_l b}{\chi} - \frac{c_l \ln(c_l + b) + b\ln(c_l + b)}{c_l + b} + \frac{\sqrt{c_l^2 + b^2}}{c_l + b} - \frac{1}{2} + 0.477 \frac{\chi}{c_l + b} \right];$$

$$L_{12}^0 = g_0^{21} = \frac{\mu_0 c_l}{\pi} \left[\ln \frac{c_l}{\chi} + \frac{1}{2} \right] + \frac{\mu_0 b}{\pi} \left[\ln \frac{b + \sqrt{b^2 + d^2}}{d} - \frac{\sqrt{b^2 + d^2}}{b} + \frac{d}{b} \right]$$
$$- \frac{\mu_0 b}{\pi} \left[\ln \frac{b + \sqrt{b^2 + (d + c_l)^2}}{d + c_l} - \frac{\sqrt{b^2 + (d + c_l)^2}}{b} + \frac{d + c_l}{b} \right],$$

$$(6.34)$$

where $d = (a - c_l)/2$. As it follows from the analysis of scheme shown in Fig. 6.32a, the mutual inductances between the coils' wires and the levitated micro-object are reduced to the analysis of the mutual inductance of the system of the parallel wires. In order to compile the terms of (3.39), let us define the mutual inductance between coil's straight wire and an element of eddy current circuit as it is shown in Fig. 6.32b. The element of eddy current circuit is highlighted in red. Using equation of mutual inductance of two parallel wires [9, p. 306] having the same length, l_c, the following auxiliary function can be defined as

$$M_a(l_c, f(x_e, x_c)) = \frac{\mu_0 l_c}{\pi} \left[\ln \frac{1 + \sqrt{1 + \xi^2}}{\xi} - \sqrt{1 + \xi^2} + \xi \right], \qquad (6.35)$$

where $\xi = f/l_c$ is the dimensionless parameter, l_c is the length of the coil wire, f is the distance between two wires, which can be calculated as

$$f(x_e, x_c) = \sqrt{(h + q_v - x_e \sin \alpha)^2 + (x_c - x_e \cos \alpha + q_l)^2}, \qquad (6.36)$$

where x_c is the coordinate of location of the coil wire along the X-axis and x_e is the coordinate of location of the element of eddy current circuit along the X'-axis (equilibrium state) as shown in Fig. 6.32b. Assuming that the displacements are small and using auxiliary function (6.35), the mutual inductance between element of the eddy current circuit and coil's wire as shown in Fig. 6.32b can be written as follows [10, p. 45]:

$$M(x_e, x_c) = (M_a((l_c + b \cos \beta)/2, f(x_e, x_c)) - M_a((l_c - b \cos \beta)/2, f(x_e, x_c))). \qquad (6.37)$$

Noting that the latter equation is derived for the case when the geometrical centres of the coil wire and element of eddy current circuit are aligned. In order to take into account, the number of windings, Eq. (6.37) can be modified as follows:

$$M_\iota(x_e, x_c) = (M_a((l_c + b \cos \beta)/2, f_\iota(x_e, x_c)) - M_a((l_c - b \cos \beta)/2, f_\iota(x_e, x_c))), \qquad (6.38)$$

where

$$f_\iota(x_e, x_c) = \sqrt{(h + p \cdot \iota + q_v - x_e \sin \alpha)^2 + (x_c - x_e \cos \alpha + q_l)^2}. \quad (6.39)$$

Hence, considering pairwise wires of coils and accounting for (6.38) terms m_0^{kj} are

$$
\begin{aligned}
m_0^{11} &= 2 \sum_{\iota=0}^{N-1} [M_\iota(c_l/2 + d, c_s/2) - M_\iota(-c_l/2 - d, c_s/2)]; \\
m_0^{22} &= 2 \sum_{\iota=0}^{M-1} [M_\iota(c_l/2, c_l/2) - M_\iota(-c_l/2, c_l/2)]; \\
m_0^{12} &= 2 \sum_{\iota=0}^{M-1} [M_\iota(c_l/2 + d, c_l/2) - M_\iota(-c_l/2 - d, c_l/2)]; \\
m_0^{21} &= 2 \sum_{\iota=0}^{N-1} [M_\iota(c_l/2, c_s/2) - M_\iota(-c_l/2, c_s/2)].
\end{aligned}
\quad (6.40)
$$

For deriving derivatives of (6.38) with respect generalized coordinates q_v, q_l and α, a general rule can be used such as

$$
\begin{aligned}
\frac{\partial M_a}{\partial q} &= \frac{\partial M_a}{\partial \xi} \frac{\partial \xi}{\partial q}; \\
\frac{\partial^2 M_a}{\partial q^2} &= \frac{\partial^2 M_a}{\partial \xi^2} \left(\frac{\partial \xi}{\partial q}\right)^2 + \frac{\partial M_a}{\partial \xi} \frac{\partial^2 \xi}{\partial q^2}.
\end{aligned}
\quad (6.41)
$$

The ξ - derivatives of M_a are

$$
\begin{aligned}
\frac{\partial M_a}{\partial \xi} &= \frac{\mu_0 l_c}{\pi} \left[1 - \frac{1}{\xi} - \frac{\xi}{1 + \sqrt{1 + \xi^2}} \right]; \\
\frac{\partial^2 M_a}{\partial \xi^2} &= \frac{\mu_0 l_c}{\pi} \left[\frac{1}{\xi^2} - \frac{1}{(1 + \sqrt{1 + \xi^2})\sqrt{1 + \xi^2}} \right].
\end{aligned}
\quad (6.42)
$$

The derivatives of ξ with respect to q_v at the equilibrium point:

$$
\begin{aligned}
\frac{\partial \xi}{\partial q_v} &= \frac{1}{l_c} \frac{h}{\sqrt{h^2 + (x_c - x_e)^2}}; \\
\frac{\partial^2 \xi}{\partial q_v^2} &= \frac{1}{l_c} \frac{(x_c - x_e)^2}{\sqrt[3]{h^2 + (x_c - x_e)^2}}.
\end{aligned}
\quad (6.43)
$$

The derivatives of ξ with respect to q_l at the equilibrium point:

$$
\begin{aligned}
\frac{\partial \xi}{\partial q_l} &= \frac{1}{l_c} \frac{x_c - x_e}{\sqrt{h^2 + (x_c - x_e)^2}}; \\
\frac{\partial^2 \xi}{\partial q_l^2} &= \frac{1}{l_c} \frac{h^2}{\sqrt[3]{h^2 + (x_c - x_e)^2}}.
\end{aligned}
\quad (6.44)
$$

The derivatives of ξ with respect to α at the equilibrium point:

$$\frac{\partial \xi}{\partial \alpha} = \frac{1}{l_c} \frac{x_e h}{\sqrt{h^2 + (x_c - x_e)^2}};$$
$$\frac{\partial^2 \xi}{\partial \alpha^2} = \frac{1}{l_c} \frac{x_e^2 (x_c - x_e)^2}{\sqrt[3]{h^2 + (x_c - x_e)^2}}.$$

(6.45)

The cross-derivatives of ξ are

$$\frac{\partial^2 \xi}{\partial q_v \partial \alpha} = \frac{1}{l_c} \frac{x_e (x_c - x_e)^2}{\sqrt[3]{h^2 + (x_c - x_e)^2}};$$
$$\frac{\partial^2 \xi}{\partial q_l \partial \alpha} = -\frac{1}{l_c} \frac{x_e h (x_c - x_e)}{\sqrt[3]{h^2 + (x_c - x_e)^2}};$$
$$\frac{\partial^2 \xi}{\partial q_l \partial q_v} = -\frac{1}{l_c} \frac{h (x_c - x_e)}{\sqrt[3]{h^2 + (x_c - x_e)^2}}.$$

(6.46)

For generalized coordinates q_v, q_l and α, the derivative of $M_\iota(x_e, x_c)$ with respect to these coordinates has the following general form:

$$\frac{\partial M_{(x_e, x_c), \iota}}{\partial q} = \frac{\partial M_a}{\partial \xi'_\iota} \frac{\partial \xi'_\iota}{\partial q} - \frac{\partial M_a}{\partial \xi''_\iota} \frac{\partial \xi''_\iota}{\partial q},$$

(6.47)

where $\xi'_\iota = 2 f_\iota(x_e, x_c)/(l_c + b)$ and $\xi''_\iota = 2 f_\iota(x_e, x_c)/(l_c - b)$. Let us consider separately derivative of M_ι with respect to β. Starting with the estimation of derivative of ξ'_ι with respect to β at the equilibrium point, we have

$$\frac{\partial \xi'_\iota}{\partial \beta} = 0;$$
$$\frac{\partial^2 \xi'_\iota}{\partial \beta^2} = 2 f_\iota(x_e, x_c) \frac{b}{(l_c + b)^2}.$$

(6.48)

Accounting for the later equations, the first and the second β-derivatives of $M_a((l_c + b \cos \beta)/2, f_\iota(x_e, x_c))$ for equilibrium point can be written as

$$\frac{\partial M_a}{\partial \beta} = 0;$$
$$\frac{\partial^2 M_a}{\partial \beta^2} = \frac{\partial M_a}{\partial \xi'_\iota} \frac{\partial^2 \xi'_\iota}{\partial \beta^2} - \frac{\mu_0 b}{2\pi} \left[\ln \frac{1 + \sqrt{1 + \xi'^2_\iota}}{\xi'_\iota} - \sqrt{1 + \xi'^2_\iota} + \xi'_\iota \right].$$

(6.49)

For terms m_l^{kj}, we can write

$$m_v^{11} = 2 \sum_{\iota=0}^{N-1} \left[\frac{\partial M_{(c_l/2+d,c_s/2),\iota}}{\partial q} - \frac{\partial M_{(-c_l/2-d,c_s/2),\iota}}{\partial q} \right];$$

$$m_v^{22} = 2 \sum_{\iota=0}^{M-1} \left[\frac{\partial M_{(c_l/2,c_l/2),\iota}}{\partial q} - \frac{\partial M_{(-c_l/2,c_l/2),\iota}}{\partial q} \right];$$

$$m_v^{12} = 2 \sum_{\iota=0}^{M-1} \left[\frac{\partial M_{(c_l/2+d,c_l/2),\iota}}{\partial q} - \frac{\partial M_{(-c_l/2-d,c_l/2),\iota}}{\partial q} \right]; \qquad (6.50)$$

$$m_v^{21} = 2 \sum_{\iota=0}^{N-1} \left[\frac{\partial M_{(c_l/2,c_s/2),\iota}}{\partial q} - \frac{\partial M_{(-c_l/2,c_s/2),\iota}}{\partial q} \right];$$

$$m_l^{11} = m_l^{22} = m_l^{12} = m_l^{21} = 0;$$
$$m_\alpha^{11} = m_\alpha^{22} = m_\alpha^{12} = m_\alpha^{21} = 0;$$
$$m_\beta^{11} = m_\beta^{22} = m_\beta^{12} = m_\beta^{21} = 0.$$

Terms m_{ll}^{kj} can be written as follows. For generalized coordinate, q_l, we have

$$m_{ll}^{11} = 2 \sum_{\iota=0}^{N-1} \left[\frac{\partial^2 M_{(c_l/2+d,c_s/2),\iota}}{\partial q_l^2} - \frac{\partial^2 M_{(-c_l/2-d,c_s/2),\iota}}{\partial q_l^2} \right];$$

$$m_{ll}^{12} = 2 \sum_{\iota=0}^{M-1} \left[\frac{\partial^2 M_{(c_l/2+d,c_l/2),\iota}}{\partial q_l^2} - \frac{\partial^2 M_{(-c_l/2-d,c_l/2),\iota}}{\partial q_l^2} \right]; \qquad (6.51)$$

$$m_{ll}^{22} = 0; \ m_{ll}^{21} = 0.$$

For other generalized coordinates q_v, α and β, the second derivatives can be found by using the general rule given below:

$$m_{qq}^{11} = 2 \sum_{\iota=0}^{N-1} \left[\frac{\partial^2 M_{(c_l/2+d,c_s/2),\iota}}{\partial q^2} - \frac{\partial^2 M_{(-c_l/2-d,c_s/2),\iota}}{\partial q^2} \right];$$

$$m_{qq}^{12} = 2 \sum_{\iota=0}^{M-1} \left[\frac{\partial^2 M_{(c_l/2+d,c_l/2),\iota}}{\partial q^2} - \frac{\partial^2 M_{(-c_l/2-d,c_l/2),\iota}}{\partial q^2} \right];$$

$$m_{qq}^{22} = 2 \sum_{\iota=0}^{M-1} \left[\frac{\partial^2 M_{(c_l/2,c_l/2),\iota}}{\partial q^2} - \frac{\partial^2 M_{(-c_l/2,c_l/2),\iota}}{\partial q^2} \right]; \qquad (6.52)$$

$$m_{qq}^{21} = 2 \sum_{\iota=0}^{M-1} \left[\frac{\partial^2 M_{(c_l/2,c_s/2),\iota}}{\partial q^2} - \frac{\partial^2 M_{(-c_l/2,c_s/2),\iota}}{\partial q^2} \right].$$

Hence, the determinants Δ and Δ_{ks} can be defined.

The structure of the model is given in Table 3.1; a condition for the stable levitation in air becomes as follows:

$$\begin{cases} c_{vv} > 0; \ c_{ll} > 0; \ c_{\alpha\alpha} > 0; \ c_{\beta\beta} > 0; & (6.53a) \\ c_{ll} \cdot c_{\alpha\alpha} > c_{l\alpha}^2. & (6.53b) \end{cases}$$

Table 6.6 Parameters of the prototype of micro-transporter

The levitation coil width, c_l^1	1500 mm
The stabilization coil width, c_s^1	2900 mm
The coils pitch of winding, p	25 mm
Number of windings for stabilization coil, N	10
Number of windings for levitation coil, M	14
Length of the track, l_c	7000 mm
Length of micro-object, b	2400 mm

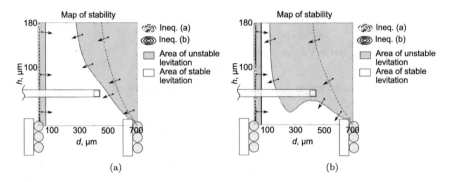

(a)   (b)

Fig. 6.33 Map of stable levitation of the prototype 3D micro-transporter: **a** the stability map for the length of the micro-object: $b = 2.8$ mm; **b** for the length $b = 0.5$ mm

Geometrical[1] parameters of the transporter prototype are defined by the schematic shown in Fig. 6.32a. The parameters of this particular prototype are presented in Table 6.6. Considering a current of 120 mA in each coil and a phase shift of 180°, the map of stability in terms of levitation height h and width $d = (a - c_s)/2$ is shown in Fig. 6.33. The figure shows two cases, namely, when the length of the micro-object is $b = 2.8$ mm and $b = 0.5$ mm. In general, the analysis of the map indicates that stable levitation in this prototype is possible for a rectangular-shaped micro-object with a length of 2.8 mm, when the width is within the range from 1.7 to 2.8 mm. The experimental study for the square-shaped micro-object having a size of 1.5 mm proves the fact that, for a micro-object with a width less than 1.7 mm, stable levitation is not possible. Another important feature which is reflected by this approach is that decreasing the length of a micro-object leads to a decrease in the area of stable levitation, and for a particular value of the length (in this case: $b = 0.5$ mm), stable levitation for any width is impossible. This fact was verified experimentally and agrees well with the theory as shown in Fig. 6.33b.

---

[1]Parameter is defined in Fig. 6.32a.

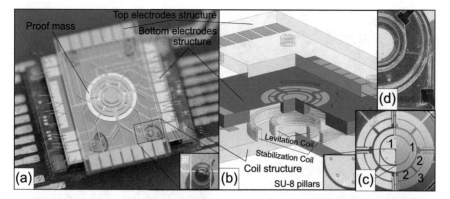

Fig. 6.34 The sensor: **a** the prototype glued to a PCB. The scaled-up image at the bottom right corner shows the alignment and the SU-8 post for spacing (the top electrode structure is not connected); **b** the exploded view; **c** The electrodes patterned at the bottom (right) and top (left). Electrode structures: 1—generating negative stiffness, 2— sensing displacement, 3—feedback electrodes; **d** a view of the aligning electrode and coil structures from the rear (pyrex glass) of the device

6.3 Micro-Accelerometers

In this section, the fabrication process, as well as the operating principle of the accelerometer based on adjustable spring constant, including its necessary service electronics for signal processing and preliminary experimental results are discussed.

6.3.1 Fabrication

The sensor consists of three structures fabricated independently at the wafer scale, namely, a coil structure, and the upper and lower electrode structures. These structures were aligned and assembled into a sandwich by flip-chip bonding into one device with the dimensions: $9.2 \times 9.2 \times 1.74$ mm as shown in Fig. 6.34. The coil structure consists of two coaxial 3D wire-bonded micro-coils, namely, a stabilization and levitation coil, fabricated on a Pyrex substrate using SU-8 2150, see Fig. 6.34b. For this particular device, the height of the coils is 600 μm, and the number of windings is 20 and 12 for the levitation and stabilization coil, respectively, which allow us to stably levitate an aluminium disc-shaped proof mass with a diameter of 3.2 mm and thickness of 30 μm at a levitation height of 150 μm. The bottom electrode structure was fabricated on an SOI wafer having a device layer of 40 μm, a buried oxide of 2 μm and a handling layer of 600 μm The resistivity of the silicon layer is in the range of $1-30$ ohm cm. Also, the device layer has a 500 nm oxide layer for passivation, on top of which electrodes are patterned by UV lithography of evaporated Cr/Au layers (20/50 nm), as shown in the left part of Fig. 6.34c. The SU-8 pillars cover the electrodes in order to insulate the proof mass and electrodes and reduce the contact

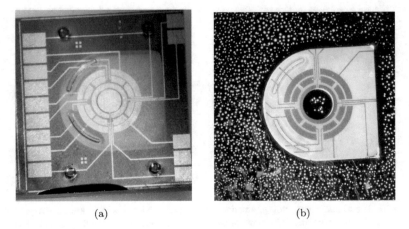

(a)　　　　　　　　　　　　　　　　(b)

Fig. 6.35 The bottom electrode structure fabricated by using a Si wafer with an SU-8 layer of 30 μm thickness: **a** The front side of the structure; **b** The rear of the structure

area between the proof mass and the surface, where the proof mass is initially lying flat. The scaled-up image at the left of Fig. 6.34c shows the SU-8 pillar having a diameter of 50 μm and a height of 10 μm. After etching the handle layer up to the buried oxide by DRIE, the bottom electrode structure was aligned and bonded onto the coil structure, as shown in Fig. 6.34d.

The top electrode structure was fabricated on a Pyrex substrate. The electrodes patterned by UV lithography of evaporated Cr/Au layers (20/150 nm) as shown in the right part of Fig. 6.34c have the same design as those on the bottom structure. To create a gap between the top and bottom structures, four SU-8 posts of 130 μm height were fabricated on the top electrode structure. Then, the top structure was aligned and bonded to the bottom one as shown in Fig. 6.34a.

To avoid using an SOI wafer, and thus also to decrease the amount of parallel capacitance arising in the patterned electrodes due to the conductivity of Si, an alternative fabrication route for the bottom electrode structure was explored based on an intrinsically doped Si wafer of 500 μm thickness with a 1 μm oxide layer for passivation, as shown in Fig. 6.35. On one side of the Si wafer, an SU-8 layer of 30–40 μm thickness was fabricated by using the epoxy resist SU-8 3025. Then, the electrodes were patterned on this SU-8 layer by UV lithography as shown in Fig. 6.35a. Instead of evaporation, the seed layers Cr/Au (20/150 nm) were sputtered on top of the SU-8 layer. Finally, etching the Si wafer through to the SU-8 layer by DRIE, a cavity for the micro-coils was fabricated as shown in Fig. 6.35b.

6.3.2 Operating Principle

The proof mass is levitated between the electrode structures. A potential U is applied to the top and bottom electrodes (denoted by the number "1") and generates an elec-

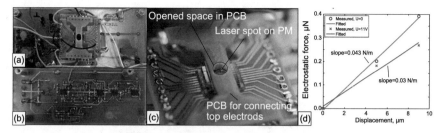

Fig. 6.36 The prototype under experimental test (this is a link on YouTube video: https://youtu.be/ewCu1i9Q5u8): **a** The device is fixed on a PCB (front side); **b** The interfacial electronics (rear side); **c** Top, bottom, and coil structures are connected to the PCB (scaled image); **d** Measurements of force against displacement

trostatic field (see Fig. 6.34c), which causes a decrease of stiffness [11]. The series of electrodes numbered "2" are patterned to realize a differential capacitance for sensing the linear displacement of the proof mass along the vertical axis. Electrodes numbered with "3" generate the electrostatic feedback-force needed to operate in a force-rebalance mode. Thus, the prototype can be considered as a levitated micro-accelerometer operating in the vertical direction and providing an adjustable positional stiffness within closed-loop control.

6.3.3 Preliminary Experimental Results

To provide a proof-of-concept and to demonstrate the successful levitation of the proof mass within the electrostatic field generated by electrodes "1", a preliminary experimental study has been performed as shown in Fig. 6.36.

Using the four pairs of electrodes labelled "2" (see Fig. 6.34c), the capacitive sensing for the vertical displacement of the proof mass, based on a capacitance half-bridge and synchronous amplitude demodulation, was implemented by using the four pairs of electrodes labelled "2" (see Fig. 6.34c). Each electrode of the pairs "2" was excited by ac voltage having an amplitude of 3 V at a frequency of 100 kHz. After traversing a charge amplifier based on OPA2107AU, the output signal was demodulated by applying a synchronous amplitude modulated signal. Using switches (ADG441) controlled by a comparator (AD8561), which in turn is synchronized with the excitation voltage, and an amplifier (OPA2107AU), the signal traverses through a mixer. The output from the mixer, passing through an instrumentation amplifier, yielded a differential signal and provided information about the linear displacement.

Coils were fed with a square wave ac current provided by a current amplifier (LCF A093R). The amplitude and frequency of the current in the coils were controlled by a function generator (Arbstudio 1104D) via a computer. A PCB for connecting the top electrodes was fabricated in such way as to leave clear the front of the levitation chip's electrodes so that a laser beam could reach the proof mass without

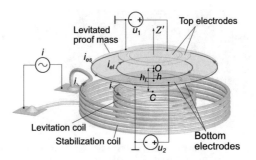

Fig. 6.37 Schematic diagram for modelling the accelerometer: **a** u_1 and u_2 are the potentials applied to the top and bottom electrodes, respectively, h is the space between an electrode's plane and the equilibrium point of the proof mass, h_l is the levitation height between the plane formed by the upper turn of the coils and the equilibrium point of the proof mass, i_{el} and i_{es} are the eddy currents corresponding to the maximum current density

Table 6.7 Parameters of the prototype, and experimental results

Parameters of the prototype		
Diameter of the proof mass	[mm]	3.2
Thickness of the proof mass	[μm]	30
Levitation height	[μm]	150
Spacing	[μm]	50
Results of measurements		
Stiffness ($U = 0$)	[N/m]	0.043
Stiffness ($U = 11V$)	[N/m]	0.03

obstruction, as shown in Fig. 6.36c. This provided us with an additional means to control the linear displacement of the levitated proof mass using a laser distance sensor (LK-G32 with a resolution of 10 nm, and a way to characterize the performance of the capacitive sensing circuit. By applying an electrostatic force generated by the electrodes "3" to the bottom surface of the proof mass, a plot of force against displacement was recorded. From the analysis of the plot, the effective suspension stiffness was estimated.

Assuming that the resulting electrostatic force was applied to the centre of the proof mass, and accounting for the area of electrode "3" of 4.3×10^{-7} m², the electrostatic force generated by the four electrodes was calculated from $F = \varepsilon_0 \varepsilon_r A / 2 \cdot (U/h)^2$, where $\varepsilon_0 = 8.85 \times 10^{-12}$ F/m is the vacuum permittivity, ε_r is the relative permittivity (for air $\varepsilon_r \approx 1$), and h is the space between an electrode's plane and the equilibrium point of the proof mass (Fig. 6.37).

The results of measurements corresponding to two cases, namely, when there is no applied electrical potential to electrodes "1", and when electrodes "1" are energized, are shown in Fig. 6.36d. Firstly, upon energizing electrodes "1", the proof mass was stably levitated and a decrease of the stiffness from 0.043 to 0.03 N/m was observed (also see Table 6.7). Secondly, a negative stiffness generated by electrodes "1" can be calculated [12] from $NS = -(\varepsilon_0 \varepsilon_r A_E U^2)/h^3$, where $A_E = 8 \times 10^{-7}$ m^2, to give $NS \approx -0.01$ N/m. This agrees well with the difference of the two measurements. The results are summarized in Table 6.7.

6.3.4 Analytical Model

A schematic diagram for modelling the accelerometer is shown in Fig. 7.1. A typical two-coil stabilization and levitation scheme, arranged to provide stable levitation of a disc-shaped proof mass, is considered. The proof mass is magnetically levitated within the static electric field generated by the top and bottom electrodes. In the general case, it is assumed that the potentials which are applied to the top and bottom electrodes, are different and denoted as u_1 and u_2, respectively, as shown in Fig. 7.1. The equilibrium point coincides with the origin O, which lies on the Z' axis of symmetry. The location of the origin is characterized by the following parameters: h is the spacing between the bottom electrode's plane and the origin, and h_l is the levitation height estimated as the distance between the plane formed by the upper turn of the coils and the origin.

The behaviour of an inductively levitated disc-shaped proof mass within the static electric field generated by the system of electrodes is strongly non-linear, described by the set of Maxwell equations. However, taking into account the fact that the induced eddy current density within the proof mass is distributed continuously but not homogenously, two circuits having maximum values of eddy current density can be identified as the representative circuit for the induced eddy current pattern as it has been discussed above. Also, assuming the quasi-static behaviour of the levitated proof mass, a simplification in the mathematical description of the sensor can be obtained. Applying the qualitative technique presented in Chap. 3, an analytical model of the accelerometer is formulated. Since its design is axially symmetric [13], the mechanical part can be represented by the three generalized coordinates, namely, q_v, q_l and θ representing vertical, lateral and angular displacements of the levitated disc, respectively, as introduced in Fig. 3.4b. Considering the capacitors as planar, and accounting for $\theta = \alpha + \beta$, the set describing the motion of the hybrid suspended proof mass becomes

$$\begin{cases} \dfrac{\partial W_e}{\partial e_1} + \dfrac{\partial \Psi}{\partial \dot{e}_1} = u_1; \quad \dfrac{\partial W_e}{\partial e_2} + \dfrac{\partial \Psi}{\partial \dot{e}_2} = u_2; \\[2mm] m\ddot{q}_v + \mu_v \dot{q}_v + mg - \dfrac{\partial(W_m - W_e)}{\partial q_v} = F_v; \\[2mm] m\ddot{q}_l + \mu_l \dot{q}_l - \dfrac{\partial(W_m - W_e)}{\partial q_l} = F_l; \\[2mm] J\ddot{\theta} + \mu_\theta \dot{\theta} - \dfrac{\partial(W_m - W_e)}{\partial \theta} = T_\theta, \end{cases} \qquad (6.54)$$

where m is the mass, J is the moment of inertia about the axis perpendicular to the disc plane and passing through the centre of mass, μ_l, μ_v and μ_θ are the damping coefficients corresponding to the appropriate generalized coordinates, g is the gravity acceleration, F_l, F_v and T_θ are the generalized forces and torque corresponding to the appropriate generalized coordinates, W_m and W_e are energies stored in the magnetic and electric fields, respectively, Ψ is the dissipation function of the system, and e_1 and e_2 are the charges on the top and bottom electrodes, respectively.

A necessary but not sufficient condition for stable levitation of the proof mass, near its equilibrium point, is that the second derivatives of electromagnetic energy stored in the system, defined by the following constants $c_{ij} = -\partial^2(W_m - W_e)/\partial q_i \partial q_j$, where $i = v, l, \theta$ and $j = v, l, \theta$, must correspond to a positive definite quadratic form. Note that the necessary and sufficient conditions for stable levitation in micromachined levitation inductive systems require, in addition, to take into account the *nonconservative positional* force due to the resistivity of the proof mass, and the *dissipative* force acting on the levitated proof mass. Thus, the nonlinear set of Eq. (6.54) form a generalized analytical model of the hybrid contactless accelerometer and provides opportunities for modelling its dynamics and stability.

6.3.5   The Accelerometer Equation of Motion

In the framework of the proposed application of the device as an accelerometer, as considered in Sect. 6.3.2, the behaviour of the proof mass along the vertical direction in the hybrid levitation system is of special interest and studied in detail below. The static and dynamic responses of the device along this direction are therefore investigated.

Note that the results of the analysis of the magnetic field and corresponding gradient shown in Fig. 6.10 is applicable also for the accelerometer because of similar coils designs. Due to this fact, the force interaction along the vertical direction is reduced to an interaction between eddy current i_{el} and the levitation coil current. Neglecting the generalized coordinates q_l and q_θ, and also assuming that the resistivity of the conducting proof mass, and its linear and angular velocities are small, no damping exists, and $u_1 = u_2 = U$, then the exact quasi-static nonlinear model, which describes the behaviour of the proof mass along the vertical axis, is

$$m\frac{d^2 q_v}{dt^2} + mg + \frac{I^2}{L}\frac{dM}{dq_v}M - \frac{A}{4}\frac{U^2}{(h - q_v)^2} + \frac{A}{4}\frac{U^2}{(h + q_v)^2} = F_v, \qquad (6.55)$$

where I is the amplitude of a harmonic current i in the coils, L is the self-inductance of the proof mass, M is the mutual inductance between the proof mass and coils, and U is the applied voltage to the electrodes. Each electrode set has the same area of A_e, $A = \varepsilon_0 A_e$, where ε_0 is the permeability of free space.

In the general case, the mutual inductance M is a complex analytical function. This represents the main difficulty for the analytical study of the accelerometer model (6.55). However, applying the same reasoning as it was discussed in Sect. 6.1.4, firstly, levitation coil and eddy current circuit can be considered as filamentary circles. Hence, the mutual inductance between the levitation coil and eddy current can be described by the Maxwell formula (see Sect. 4.3), thus the Eq. (6.4) can be used. Then, accounting for (6.4), model (6.55) of accelerometer becomes

$$m\frac{d^2q_v}{dt^2} + mg - \frac{I^2a^2}{L}\left[\left(\frac{2}{k} - k\right)K(k) - \frac{2}{k}E(k)\right]\frac{2}{k^2}$$
$$\times \left[\frac{2 - k^2}{2(1 - k^2)}E(k) - K(k)\right] \cdot \frac{\xi^2(1 + \frac{q_v}{h_l})}{h_l(1 + \xi^2(1 + \frac{q_v}{h_l})^2)^{3/2}} - \frac{AU^2q_v}{(h - q_v^2)^2} = F_v,$$

$$(6.56)$$

where $a = r_l\mu_0$ and $\xi = h_l/(2r_l)$. Model (6.56) is analytical, nonlinear, and quasi-exact, but due to the elliptic integrals, it can be studied only numerically. For further analysis, model (6.56) is presented in dimensionless form as follows:

$$\frac{d^2\lambda}{d\tau^2} + 1 - \eta\left[\left(\frac{2}{k} - k\right)K(k) - \frac{2}{k}E(k)\right]\frac{2}{k^2}$$
$$\times \left[\frac{2 - k^2}{2(1 - k^2)}E(k) - K(k)\right] \cdot \frac{\xi^2(1 + \lambda)\kappa}{(1 + \xi^2(1 + \lambda)^2)^{3/2}} - \frac{\beta\lambda}{(1 - \lambda^2)^2} = \tilde{F},$$

$$(6.57)$$

where $\tau = \sqrt{g/ht}$, $\lambda = q_v/h$, $\eta = I^2a^2/(mghL)$, $\beta = AU^2/(mgh^2)$, $\kappa = h/h_l$ and $\tilde{F} = F_v/mg$. Secondly, upon ensuring that the parameter $\xi = h_l/(2r_l)$ is less than 0.3 as described further down, the formula (6.4) can be well approximated by the logarithmical function (see, Sect. 4.3.2):

$$M = \mu_0r_l\left[\ln\frac{4}{\xi(1 + y/h_l)} - 2\right].$$

$$(6.58)$$

Hence, taking into account the later equation, the following reduced analytical model of the sensor is proposed:

$$m\frac{d^2y}{dt^2} + mg - \frac{I^2a^2}{L}\frac{1}{h_l + y}\left[\ln\frac{4}{\xi(1 + y/h_l)} - 2\right] - \frac{AU^2q_v}{(h^2 - q_v^2)^2} = F_v. \quad (6.59)$$

In dimensionless form, Eq. (6.59) becomes

$$\frac{d^2\lambda}{d\tau^2} + 1 - \eta\frac{\kappa}{1 + \kappa\lambda}\left[\ln\frac{4}{\xi(1 + \kappa\lambda)} - 2\right] - \frac{\beta\lambda}{(1 - \lambda^2)^2} = \tilde{F}. \quad (6.60)$$

As it has been mentioned above, the accuracy of approximation of modelling the electromagnetic force in the reduced model (6.60) is depend on the parameter $\xi = h_l/(2r_l)$. Figure 6.38 shows the relative error of the approximation as a function of this parameter and its change with the dimensionless displacement, λ. Analysis

Fig. 6.38 Accuracy of modelling the electromagnetic force by means of the reduced model (6.60) as compared to the quasi-exact model (6.57)

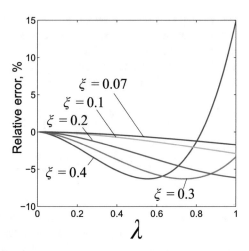

of Fig. 6.38 reveals that, if the parameter ξ is less than 0.3, the electromagnetic force is well approximated by the logarithmic function (6.58) with the relative error less than 6 %. When parameter ξ vanishes, the error between the exact Eq. (6.4) and the approximation (6.58) also vanishes. It is worth noting that, for all known prototypes of μ-HCS published in the literature, parameter ξ is less than 0.25. This fact indicates the applicability of the reduced model for further analytical study of μ-HCS. Hence, model (6.60) is the main framework for further analysis of the static and dynamic pull-in.

6.3.6 Static Pull-In Instability

The load-free behaviour of the device upon changing the strength of the electric field, characterized by the dimensionless parameter β (dimensionless squared voltage) is studied. For this reason, Eq. (6.60) is written as a set in terms of the phase coordinates:

$$\begin{cases} \dfrac{d\lambda}{d\tau} = \omega; \\ \dfrac{d\omega}{d\tau} = -1 + \eta\dfrac{\kappa}{1 + \kappa\lambda}\left[\ln\dfrac{4}{\xi(1 + \kappa\lambda)} - 2\right] + \dfrac{\beta\lambda}{(1 - \lambda^2)^2}. \end{cases} \tag{6.61}$$

From (6.61), the equilibrium state of the system can be defined as

$$f(\lambda, \beta) = -1 + \eta\dfrac{\kappa}{1 + \kappa\lambda}[D - \ln(1 + \kappa\lambda)] + \dfrac{\beta\lambda}{(1 - \lambda^2)^2}, \tag{6.62}$$

where $D = \ln\frac{4}{\xi} - 2$ is the design parameter depending on ξ. At the equilibrium point $\lambda = 0$, the function f must equal zero; This point requires that parameter $\eta = 1/D$.

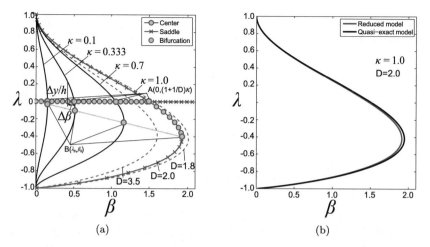

Fig. 6.39 Bifurcation diagram: **a** Dashed red lines show the evolution of the bifurcation map depending on constant D ($\kappa = 1.0$); Solid lines depict the evolution of the bifurcation map depending on spacing $\kappa = h/h_l$ ($D = 2.0$); **b** Comparison of the quasi-exact and reduced models for $D = 2.0$, $\kappa = 1.0$ and $\xi = 0.07$ (the relative error is less than 2%)

Hence, the static equilibrium state of the system, which relates the vertical coordinate with the strength of the electric field, is

$$f(\lambda, \beta) = -\frac{\kappa\lambda}{1+\kappa\lambda} - \frac{\ln(1+\kappa\lambda)}{D(1+\kappa\lambda)} + \frac{\beta\lambda}{(1-\lambda^2)^2} \equiv 0. \tag{6.63}$$

Since the vertical displacement of the proof mass is limited by the positions of the top and bottom electrodes, the variable λ is varied within a range of $-1 \le \lambda \le 1$. Also, taking into account that constant D and κ can be considered within the following ranges of $1 < D < 4.0$ and $0 < \kappa \le 1$, the bifurcation diagram, which relates the distribution of saddles (unstable equilibrium), centres (stable equilibrium) and bifurcations with the dimensionless square voltage β, is shown in Fig. 6.39a. Note that the presented prototype of hybrid levitation accelerometer presented in Sect. 6.1.2 has the following values of dimensionless parameters, namely, $D = 2.0456$, $\xi = 0.07$ and $\kappa = 0.3333$. Analysis of the diagram shows that two bifurcation points can be recognized, denoted as A and B. Both bifurcation points correspond to the pull-in instability. This means that, once the strength of electric field has achieved the value characterized by β_A, the proof mass is pulled in and moves towards the top electrodes. At point B, where the strength of electric field is characterized by β_B, the proof mass at the position characterized by λ_B is also pulled in but moves already towards the bottom electrodes.

The bifurcation point A is defined by the following parameters:

$$\lambda_A = 0; \quad \beta_A = \kappa(1+1/D). \tag{6.64}$$

Parameters β_B and λ_B characterizing bifurcation point B (static pull-in instability) are defined numerically as the solution of the following set of equations:

$$
\begin{cases}
-3D\kappa^2\lambda^4 - \kappa(4D+1)\lambda^3 - \kappa^2 D\lambda^2 + \kappa\lambda - (2\kappa\lambda^3 + 3\lambda^2 + 2\kappa\lambda + 1)\ln(1+\kappa\lambda) = 0; \\
\beta = \left(\kappa\lambda + \dfrac{\ln(1+\kappa\lambda)}{D}\right)\dfrac{(1-\lambda^2)}{\lambda(1+\kappa\lambda)}.
\end{cases}
\tag{6.65}
$$

When κ is small, then the set (6.65) has an approximate solution

$$
\lambda_B \approx -\frac{\kappa}{4}\frac{D+3/2}{D+1}; \quad \beta_B \approx \kappa(1+1/D)\left(1+\frac{\kappa^2}{4}\frac{D+3/2}{D+1}\right).
\tag{6.66}
$$

In addition, a comparison between reduced model (6.60) and the quasi-exact model (6.57) is performed for the considered design of the sensor in this work, characterized by the following dimensionless parameters $D = 2$ and $\xi = 0.07$. The result of this comparison is presented in Fig. 6.39b. Analysis of Fig. 6.39b reveals that the relative error is not in excess of 2%.

The ranges of parameters $\lambda = 0, 0 \leq \beta < \beta_A$ and $-\lambda_B < \lambda < 0, \beta_A \leq \beta < \beta_B$, establish a stable state of equilibrium (see Fig. 6.39a). A region near the bifurcation point A is of special interest because it defines a state of zero stiffness of the suspension. As seen, a decrease of stiffness leads to decreasing a range of linear displacement of the proof mass. Near bifurcation point A, the range of displacement becomes

$$
h\frac{\Delta\beta}{\kappa(1+1/D)} = h\frac{\kappa(1+1/D) - \beta}{\kappa(1+1/D)} \geq \Delta y.
\tag{6.67}
$$

Using Eq. (6.67), the minimum possible value of linear stiffness still capable of upholding stable levitation can be estimated. For instance, in the fabricated design of the sensor (see Table 6.7), upon controlling the linear displacement of the proof mass Δy within a range of ± 1 µm, the relative minimization of the stiffness can be expected to be around 0.007. It means that the initial stiffness generated by the inductive suspension can be reduced by two orders of magnitude. Note that the design of the sensor corresponds to a bifurcation curve with $\kappa = 0.3333$ as shown in Fig. 6.39a.

6.3.7   Dynamic Pull-In Instability

An equation for the integral curves of set (6.61) can be obtained as follows:

$$
\frac{d\omega}{d\lambda} = \frac{-\dfrac{\kappa\lambda}{1+\kappa\lambda} - \dfrac{\ln(1+\kappa\lambda)}{D(1+\kappa\lambda)} + \dfrac{\beta\lambda}{(1-\lambda^2)^2}}{\omega}.
\tag{6.68}
$$

Fig. 6.40 Static and dynamic bifurcation diagrams: solid black lines correspond to unphysical stagnation

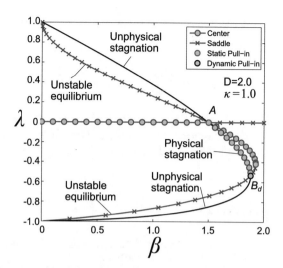

Integrating (6.68), the equation of energy is obtained as

$$\omega^2 + 2\lambda - 2\frac{\ln(1 + \kappa\lambda)}{\kappa} + \frac{\ln^2(1 + \kappa\lambda)}{D\kappa} - \frac{\beta}{1 - \lambda^2} = G, \qquad (6.69)$$

where G is an arbitrary constant of the integration. From analysis of (6.69), it is very important to note the following observation that to operate the device properly, it is required to remove the energy of the electric field characterized by parameter β from the system, to satisfy the initial conditions. Since G is an arbitrary constant, it can be chosen to be equal to $-\beta$. Then, the final form of the integral equation becomes

$$\omega^2 + 2\lambda - 2\frac{\ln(1 + \kappa\lambda)}{\kappa} + \frac{\ln^2(1 + \kappa\lambda)}{D\kappa} - \frac{\beta\lambda^2}{1 - \lambda^2} = G'. \qquad (6.70)$$

From (6.70), the dynamic equilibrium state can be written as

$$f_d(\lambda, \beta) = 2\lambda - 2\frac{\ln(1 + \kappa\lambda)}{\kappa} + \frac{\ln^2(1 + \kappa\lambda)}{D\kappa} - \frac{\beta\lambda^2}{1 - \lambda^2} \equiv 0. \qquad (6.71)$$

Using (6.71), the bifurcation diagram can be plotted as shown in Fig. 6.40. Similar to the static bifurcation diagram, it has two pull-in instability points (see Fig. 6.40). One point has the same coordinates as point A shown in (6.64) corresponding the static pull-in instability, but B_d has different coordinates compared to the static pull-in point B, and can be found by numerically solving the following set:

$$\begin{cases} \dfrac{2\kappa\lambda^2(1 - \lambda^2)}{1 + \kappa\lambda} - 4\lambda + \left[\dfrac{\kappa\lambda(1 - \lambda^2)}{1 + \kappa\lambda} + 2D - \ln(1 + \kappa\lambda)\right]\dfrac{2}{\kappa D}\ln(1 + \kappa\lambda) = 0; \\[4mm] \beta = \left(2\lambda - 2\dfrac{\ln(1 + \kappa\lambda)}{\kappa} + \dfrac{\ln^2(1 + \kappa\lambda)}{D\kappa}\right)\dfrac{(1 - \lambda^2)}{\lambda^2}. \end{cases}$$

$$(6.72)$$

Similar to Sect. 6.3.6, we consider the case when κ is small, then the set (6.72) has an approximate solution

$$\lambda_{B_d} \approx -\kappa \frac{D/3 + 1/2}{D + 1}; \quad \beta_{B_d} \approx \frac{\kappa}{D} \left(1 - \kappa^2 \frac{(D/3 + 1/2)^2}{(D + 1)^2} \right). \tag{6.73}$$

Note that, when the spacing κ trends to zero, the static and dynamic pull-in displacements tend to their zero initial position and pull-in voltages also tend to zero. Once $\kappa = 0$, all static and dynamic pull-in points merge into one zero point.

References

1. O.G. Levi, N. Krakover, S. Krylov, Sub-g bistable frequency sensor with a tunable threshold, in *2019 IEEE SENSORS* (2019), pp. 1–4
2. Z. Lu, F. Jia, J. Korvink, U. Wallrabe, V. Badilita, Design optimization of an electromagnetic microlevitation System based on copper wirebonded coils, in *2012 Power MEMS* (Atlanta, GA, USA 2012), pp. 363–366
3. K.V. Poletkin, J.G. Korvink, Efficient calculation of the mutual inductance of arbitrarily oriented circular filaments via a generalisation of the Kalantarov-Zeitlin method. J. Magn. Magn. Mater. **483**, 10–20 (2019). https://doi.org/10.1016/j.jmmm.2019.03.078
4. A.A. Andronov, A.A. Vitt, S.E. Khaikin, *Theory of Oscillators: Adiwes International Series in Physics*, vol. 4 (Elsevier, 2013)
5. D. Elata, H. Bamberger, On the dynamic pull-in of electrostatic actuators with multiple degrees of freedom and multiple voltage sources. J. Microelectromechanical syst. **15**(1), 131–140 (2006)
6. K. Poletkin, Z. Lu, U. Wallrabe, V. Badilita, A new hybrid micromachined contactless suspension with linear and angular positioning and adjustable dynamics. J. Microelectromechanical Syst. **24**(5), 1248–1250 (2015). https://doi.org/10.1109/JMEMS.2015.2469211
7. I. Sari, M. Kraft, A MEMS Linear Accelerator for Levitated Micro-objects. Sensor. Actuat. A-Phys. **222**, 15–23 (2015)
8. Z. Lu, *Micromachined Inductive Suspensions with 3D Wirebonded Microcoils* (Shaker Verlag, 2017)
9. E.B. Rosa, *The Self and Mutual Inductances of Linear Conductors* (US Department of Commerce and Labor, Bureau of Standards, 1908)
10. F. Grover, *Inductance Calculations: Working Formulas and Tables* (Dover publications, Chicago, 2004)
11. K.V. Poletkin, A.I. Chernomorsky, C. Shearwood, A proposal for micromachined accelerometer, base on a contactless suspension with zero spring constant. IEEE Sens. J. **12**(07), 2407–2413 (2012). https://doi.org/10.1109/JSEN.2012.2188831
12. K.V. Poletkin, A.I. Chernomorsky, C. Shearwood, A proposal for micromachined dynamically tuned gyroscope, based on contactless suspension. IEEE Sens. J. **12**(06), 2164–2171 (2012). https://doi.org/10.1109/JSEN.2011.2178020
13. K. Poletkin, Z. Lu, U. Wallrabe, J. Korvink, V. Badilita, Stable dynamics of micro-machined inductive contactless suspensions. Int. J. Mech. Sci. **131–132**, 753–766 (2017). https://doi.org/10.1016/j.ijmecsci.2017.08.016

Chapter 7
Mechanical Thermal Noise in Levitation Micro-Gyroscopes

Micromachined *levitation gyroscopes* (MLGs) employing the operating principle of a classical spinning gyroscope for sensing an angular rate have been intensively studied over recent decades. The main feature of such gyroscopes is that a spinning rotor is suspended, without physical attachment to an inertial frame by means of using, in particular, electromagnetic levitation phenomena. Many different prototypes of MLGs have been demonstrated. According to the physical principle of electromagnetic levitation implemented in the sensor, gyroscopes can be classified as inductive [1–4], diamagnetic [5, 6], electric [7–12] and hybrid MLG [13, 14]. It is worth noting that the electrostatic gyroscope reported in [12] has performance figures close to tactical grade and is already commercially available on the market.

The performance of an ideal micromachined gyroscope is restricted by mechanical thermal noise (MTN) arising in electrical and mechanical components because of energy dissipation. On the one hand, mechanical thermal noise defines the theoretical performance of a micro-gyroscope [15]. On the other hand, due to a dramatic increase in accuracy of micromachining over the past decade, the systematic errors of micro-gyroscopes caused by fabrication imperfections are already comparable with random errors induced by noise. Thus, a detailed understanding of the mechanisms of MTN is essential to improve the current performance, as well as for the further development of micromachined gyroscopes aiming for higher accuracy, corresponding, for example, to tactical and navigation tasks.

A key source of energy dissipation in mechanical elements of micro-gyroscopes is friction. Here, an attractive advantage of an MLG is that mechanical friction is controlled by vacuum quality, in contrast, for instance, with *Coriolis vibratory gyroscopes* (CVG), in which mechanical friction is defined by dissipation effects inside a constructive material. Moreover, in addition to controlled friction in an MLG, the rotor rotation speed is another important parameter, which can be varied over a wide range, from thousands to millions of rpm, providing the possibility of a further increase in the performance of the gyroscope [11]. This means that a level

K. Poletkin, *Levitation Micro-Systems*, Microsystems and Nanosystems, https://doi.org/10.1007/978-3-030-58908-0_7

of MTN in an MLG is heavily dependent on a gyroscope design. Under the same
operating conditions, mechanical thermal noise in MLGs is potentially significantly
less than in their vibrating counterparts. Consequently, MLGs potentially show a
better performance.

The preservation of angular position stiffness in all reported prototypes of MLG
[16], which can be either positive or negative, leads to an additional response of
a levitated gyroscope rotor to thermal fluctuating torques in the form of Johnson
noise. The study of this additional response shows that it has a significant effect on
the performance of an MLG and defines the resolution of the gyroscope. Using the
Ising criterion [17], an alternative qualitative manner as it is shown in the proceeding
chapter, the gyroscope resolution can be accurately estimated. Joining both Johnson
and Ising criteria, a confidence range of the gyroscope resolution can be obtained.

7.1 Model of an Ideal Levitated Two-Axis Rate Gyroscope

A levitated two-axis rate gyroscope has an operating principle similar to that of
a classical mechanical gyroscope. Hence, an angular rate of the gyroscope casing
produces a gyroscopic torque acting on the gyroscope's spinning rotor. The angular
rate is estimated by measuring the angular displacement of the spinning rotor. Let
us introduce the following coordinate frames (CFs) denoted as XYZ fixed to the
gyroscope's case and $x_r y_r z_r$ co-rotating with a speed of Ω with respect to XYZ. The
z_r and Z axes of the rotating and fixed CFs are coincident as is shown in Fig. 7.1.
It is assumed that the rotor is a rigid body which is statically as well as dynamically
perfectly balanced and does not deform. We assign the xyz CF to the rotor in such way

Fig. 7.1 Determination of
the position of the rotor with
respect to the fixed XYZ and
rotating $x_r y_r z_r$ coordinate
frames: the xyz CF (denoted
in blue) is assigned to the
principal axes of the rotor;
the $x'y'z'$ CF (denoted in
red) is assigned to the rotor
that the $x'y'$ surface is lying
on the xy one

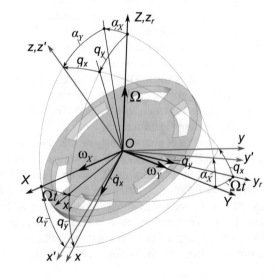

that the axes also coincide with the principle axes of rotor inertia. Assuming that in the unperturbed state (e.g., absence of a measurable angular rate) the spinning speed of the rotor has the same value of Ω, then the behaviour of an ideal gyroscope with respect to the rotating $x_r y_r z_r$ CF can be characterized by two angular generalized coordinates q_x and q_y (please refer to Fig. 7.1) and described, e.g., in [16, p. 7003] (the details of the derivation are presented in Appendix C.1)

$$\left.\begin{array}{l} J_x\ddot{q}_x + \mu_x\dot{q}_x + C_{dx}q_x - h_y\dot{q}_y = \tau_x; \\ h_x\dot{q}_x + J_y\ddot{q}_y + \mu_y\dot{q}_y + C_{dy}q_y = \tau_y, \end{array}\right\} \tag{7.1}$$

where

$$\begin{array}{l} \tau_x = \tau_Y' \sin \Omega t - \tau_X' \cos \Omega t; \\ \tau_y = \tau_Y'' \cos \Omega t + \tau_X'' \sin \Omega t \end{array} \tag{7.2}$$

are the respective torques acting on the x and y axes of the rotor, and

$$\begin{array}{l} \tau_X' = (J_z + J_x - J_y)\Omega \cdot \omega_Y + J_x\dot{\omega}_X - F_X; \\ \tau_Y' = (J_z + J_x - J_y)\Omega \cdot \omega_X - J_y\dot{\omega}_Y - F_Y; \\ \tau_X'' = (J_z - J_x + J_y)\Omega \cdot \omega_Y + J_x\dot{\omega}_X - F_X; \\ \tau_Y'' = (J_z - J_x + J_y)\Omega \cdot \omega_X - J_y\dot{\omega}_Y - F_Y \end{array} \tag{7.3}$$

are the respective torques acting on the X and Y axes, F_X and F_Y are the feedback torques acting along the given axes; $h_x = J_x(1 - \kappa_x)\Omega$; $h_y = J_y(1 - \kappa_y)\Omega$ and

$$\kappa_x = \left(J_z - J_y\right)\Big/J_x, \; \kappa_y = \left(J_z - J_x\right)\Big/J_y \tag{7.4}$$

are the respective gyroscope constructive parameters relative to the x and y axes; $C_{dx} = c_s + \kappa_x \cdot J_x \cdot \Omega^2$, $C_{dy} = c_s + \kappa_y \cdot J_y \cdot \Omega^2$ are the dynamic stiffness; and c_s is the angular position stiffness produced by a contactless suspension holding and rotating the rotor. The behaviour of the gyroscope with respect to the fixed CF is described by the angular generalized coordinates, α_X and α_Y, as shown in Fig. 7.1, which have the following relationship to q_x and q_y [18, p. 2166]

$$\left.\begin{array}{l} \alpha_X = q_x \cos \Omega t - q_y \sin \Omega t; \\ \alpha_Y = q_x \sin \Omega t + q_y \cos \Omega t. \end{array}\right\} \tag{7.5}$$

For further analysis, the rotor can be considered to be axially symmetric about the z axis, so that we have

$$J = J_x = J_y \le J_z. \tag{7.6}$$

Accounting for assumption (7.6) and introducing the complex variables $\bar{q} = q_x + jq_y$ and $\bar{\tau} = \tau_X + j\tau_Y$ ($j = \sqrt{-1}$), model (7.1) can be rewritten in a compact complex form as follows:

$$J\ddot{\bar{q}} + (\mu + jh)\dot{\bar{q}} + C_d\bar{q} = -\bar{\tau}^* e^{-j\Omega t}, \tag{7.7}$$

where the symbol $*$ denotes complex conjugation. Applying the Laplace transform to (7.7), we have

$$\left(Js^2 + (\mu + jh)s + C_d\right)\overline{q}\{s\} = -\overline{\tau}^*\{s + j\Omega\}. \tag{7.8}$$

Rewriting relationship (7.5) in terms of complex variables:

$$\overline{\alpha} = \overline{q}e^{j\Omega t}, \tag{7.9}$$

where $\overline{\alpha} = \alpha_X + j\alpha_Y$, a final linear model of an ideal two-axis levitated gyroscope, describing its behaviour with respect to the fixed CF, can be written as

$$\left(J(s - j\Omega)^2 + (\mu + jh)(s - j\Omega) + C_d\right)\overline{\alpha}\{s\} = -\overline{\tau}^*\{s\}. \tag{7.10}$$

Model (7.10) is the object of our further study.

7.2   Mechanical Thermal Noise

The main source of mechanical thermal noise in levitated micro-gyroscopes is the Brownian motion of gas molecules surrounding the spinning rotor, as well as phonons and electron gases in the constructive material of the rotor, which induce fluctuating torques and forces acting on the gyroscope rotor. In the case of thermal equilibrium between the rotor and the mentioned gases, according to the *equipartition theorem*, the average energy for each degree of freedom is equal to $k_BT/2$, where k_B is the Boltzmann constant 1.38×10^{-23} J/K and T is the absolute temperature [19, p. 128]. Since the behaviour of an ideal levitated two-axis gyroscope (7.10) is described by two generalized angular coordinates, namely, α_X and α_Y, thus, the average kinetic and potential energy accumulated by the gyroscope due to the Brownian motion of gas molecules surrounding the spinning rotor are

$$\frac{1}{2}J\langle\dot{\alpha}_X^2\rangle = \frac{1}{2}J\langle\dot{\alpha}_Y^2\rangle = \frac{1}{2}c_s\langle\alpha_X^2\rangle = \frac{1}{2}c_s\langle\alpha_Y^2\rangle = \frac{1}{2}k_BT, \tag{7.11}$$

where $\langle\cdot\rangle$ is the ensemble average. All variables in (7.11) are independent, Gaussian and have zero mean value. It is important to note that, although the gyroscope stiffness consists of two components, namely, angular position stiffness generated by a contactless suspension c_s, and dynamical stiffness due to rotation, the dynamical component of stiffness nevertheless vanishes because of the operating principle of the gyroscope [16]. As a result, (7.11) only considers the position stiffness. For the same reason, dynamical tuning [18, 20–22] in this gyroscope with an axially symmetric rotor is impossible [21, p. 45].

   According to the Nyquist relation [23], the two-sided spectral density of the fluctuating torque is

$$S_\tau(\omega) = 2k_B T \mu, \quad -\infty < \omega < \infty. \tag{7.12}$$

For the positive frequency domain, $0 < \omega < \infty$, a factor of 2 in Eq. (7.12) is replaced by a factor of 4 [24], [23, p. 112]. It is essential to study the behaviour of the gyroscope over the entire frequency domain, due to the particularities of the gyroscope model (7.10) such as its complex coefficients [25, 26].

7.3 Noise Floor. Angular Random Walk

Let us obtain the spectral density of mechanical thermal noise in the gyroscope. For this reason, model (7.10) is rewritten in the frequency domain relating the angular speeds of rotor displacements and the applied torques as follows:

$$\left[j \left(J(\omega - \Omega) - \frac{C_d}{\omega - \Omega} + h \right) + \mu \right] \dot{\bar{\alpha}} \{j\omega\} = -\bar{\tau}^* \{j\omega\}. \tag{7.13}$$

Multiplying (7.13) on its complex conjugate equation, we can write:

$$\left[\left(J(\omega - \Omega) - \frac{C_d}{\omega - \Omega} + h \right)^2 + \mu^2 \right] \dot{\bar{\alpha}} \cdot \dot{\bar{\alpha}}^* = \bar{\tau} \cdot \bar{\tau}^*. \tag{7.14}$$

The later equation provides the relationship between the square of the generalized angular speed, $\dot{\bar{\alpha}} \cdot \dot{\bar{\alpha}}^*$, and the torque, $\bar{\tau} \cdot \bar{\tau}^*$. The both vectors are directed along an axis lying on the XY surface and the equatorial surface of the rotor as shown in Fig. 7.2. As it has been noted above that the dynamic stiffness, C_d, can be either positive or negative value. Below both cases are considered.

Positive Dynamic Stiffness $C_d > 0$

Accounting for the average kinetic energy accumulated by a spinning rotor due to thermal fluctuation mentioned above, the response of the gyroscope on this fluctuation can be written as

$$\langle \tau^2 \rangle = \frac{2k_B T}{J} \left[\left(J(\omega - \Omega) - \frac{C_d}{\omega - \Omega} + h \right)^2 + \mu^2 \right]. \tag{7.15}$$

At a resonance frequency of $(\omega - \Omega) = \Omega_0 = \frac{\sqrt{4JC_d - h^2} - h}{2J}$, the mean square value has a maximum, thus

$$\langle \tau^2 \rangle = \frac{2k_B T}{J} \mu^2 = \frac{2k_B T}{Q} \mu \Omega_0, \tag{7.16}$$

where $Q = J\Omega_0/\mu$ is the quality factor. Taking into account that $Q = \Omega_0/\Delta\omega$, where $\Delta\omega$ is the bandwidth, the mean square torque Eq. (7.16) can be rewritten in a

Fig. 7.2 The determination
of position of vectors of $\bar{\tau}\bar{\tau}^*$
(green vector) and $\dot{\bar{\alpha}}\dot{\bar{\alpha}}^*$ (blue
vector)

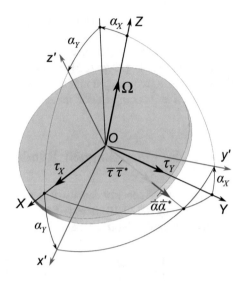

standard form as

$$\langle \tau^2 \rangle = 2k_B T \mu \Delta \omega. \tag{7.17}$$

From (7.17), the two-sided spectral density is $S_N(\omega) = 2k_B T \mu$, $-\infty < \omega < \infty$. Introducing a noise component for the measuring rate denoted as $\langle \Omega_N \rangle$ and since it produces a torque equaling $\langle \Omega_N \rangle \cdot J_z \Omega$, Hence, the steady-state spectral density corresponding to this noise component is

$$S_{\Omega_N}(\omega) = \frac{2k_B T \mu}{J_z^2 \Omega^2}, \quad -\infty < \omega < \infty. \tag{7.18}$$

For only the positive domain of frequencies, the spectral density is increased by a factor of 2, we can write:

$$S_{\Omega_N}^+(\omega) = \frac{4k_B T \mu}{J_z^2 \Omega^2}, \quad 0 < \omega < \infty. \tag{7.19}$$

Using (7.19), the angular random walk of the gyroscope becomes

$$\Omega_{ARW} = \sqrt{S_{\Omega_N}^+(\omega)} = \frac{\sqrt{4k_B T \mu}}{J_z \Omega} \frac{180 \cdot 60}{\pi}, \quad °/\sqrt{h}. \tag{7.20}$$

Negative Dynamic Stiffness $C_d < 0$

In this case, the mean square values of fluctuating torque is

Table 7.1 Parameters of levitated two-axis gyroscopes

References	Rotor					
	Shape	Size	Moments of inertia		Constructive parameter	Vacuum
			kg m^2	sec^{-1}		
		$^a d \times h \times w$	J	J_z	$1 - \kappa$	Pa
[2]	Disk	0.5×10	8.3×10^{-17}	1.7×10^{-16}	1.1×10^{-3}	–
[4]	Disk	2.2×20	6.2×10^{-14}	1.24×10^{-13}	2.2×10^{-4}	–
[8]	Ring	$4.0 \times 150 \times 300$	2.07×10^{-12}	4.15×10^{-12}	3.9×10^{-5}	0.1
[12]	Ring	$1.5 \times 50 \times 150$	2.29×10^{-14}	4.59×10^{-14}	3.3×10^{-4}	0.05
[10]	Ring	$4.0 \times 68 \times 270$	7.17×10^{-13}	1.43×10^{-12}	3.7×10^{-5}	0.6

^{a}diameter (mm) × height (μm) × width (μm)

$$\langle \tau^2 \rangle = \frac{2k_B T}{J} \left[\left(J(\omega - \Omega) + \frac{C_d}{\omega - \Omega} + h \right)^2 + \mu^2 \right]. \tag{7.21}$$

In order to find a frequency corresponding to the minimum value of mechanical impedance in Eq. (7.21), we write the $(\omega - \Omega)$-derivation of the impedance equaling zero as

$$2 \left(J(\omega - \Omega) + \frac{C_d}{\omega - \Omega} + h \right) \left(J - \frac{C_d}{(\omega - \Omega)^2} \right) = 0. \tag{7.22}$$

The solution of (7.22) shows that the mechanical impedance of Eq. (7.21) has two minimums corresponding to frequencies $(\omega - \Omega) = \Omega_0 = \pm\sqrt{C_d/J}$. Accounting for the obtained frequency $(\Omega_0 = \sqrt{C_d/J})$, the effective damping for this case can be written as follows:

$$\tilde{\mu} = \sqrt{\mu^2 + \left(h + 2\sqrt{JC_d} \right)^2}. \tag{7.23}$$

Accounting for the effective damping (7.23), the angular random walk of the gyroscope with negative dynamic stiffness is

$$\Omega_{ARW} = \frac{\sqrt{4k_B T \tilde{\mu}}}{J_z \Omega} \frac{180 \cdot 60}{\pi}, \quad °/\sqrt{h}. \tag{7.24}$$

For this case, in addition to the damping coefficient, effective damping includes also two terms, namely, gyroscope coupling coefficients and the dynamic stiffness.

Fig. 7.3 Angular random walk calculated for designs reported in [2, 4, 8, 10, 12] as a function of rotor speed in case of positive dynamic stiffness (7.20)

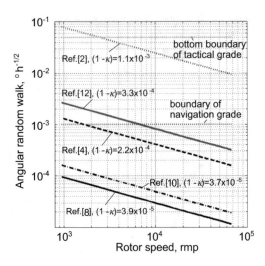

Analysis of ARW

Let us estimate the ARW of known levitated gyroscopes by means of Eq. (7.20) and (7.24), for instance, the inductive devices reported in [2, 4], and electrostatic devices reported in [8, 10, 12]. Note that the estimation is performed at room temperature (nominally $T = 300K$) for all gyroscopes mentioned above, even though the inductive gyroscopes [2, 4] had an operating temperature of more than 200°C. This fact is ignored in order to separately study the effect of rotor geometry, its inertia and speed. Moreover, the operating temperature of micromachined inductive contactless suspensions has already been reduced down to 60°C, which is not a limit, as it has been discussed in Sect. 5.2. Thus, future inductive gyroscopes will probably have operating temperatures compatible with the ambient. Table 7.1 lists the moments of inertia of the rotors and their constructive parameters, $1 - \kappa$, for the considered gyroscopes. The estimation results conducted within a range of rotor speeds from 1×10^3 to 1×10^5 rpm.

In a case of positive dynamic stiffness ($C_d > 0$), ARW in mentioned above prototypes is estimated by Eq. (7.20) and the result of estimation is shown in Fig. 7.3. Let us note that although in electrostatic gyroscopes position stiffness is negative, however, the dynamical stiffness can be still positive if the stiffness due to rotation ($\kappa J \Omega^2$) is dominated, thus

$$C_d = \kappa J \Omega^2 - c_s > 0. \qquad (7.25)$$

Upon holding (7.25), ARW in the electrostatic gyroscopes is described by Eq. (7.20). An analysis of Fig. 7.3 reveals that increasing the moments of inertia and rotor speed decreases ARW.

In a case of negative dynamic stiffness ($C_d < 0$), which is possible only in the electrostatic gyroscopes having negative angular position stiffness, ARW is calcu-

Fig. 7.4 Angular random walk calculated for electrostatic gyroscopes having negative position stiffness reported in [8, 10, 12]

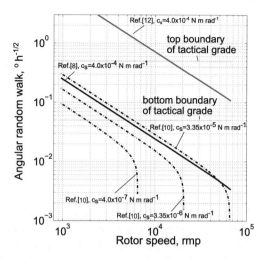

lated by (7.24) and a result of the calculation is shown in Fig. 7.4, also providing values of position stiffness for each calculation (the values of position stiffness are shown in absolute dimension since the sign of stiffness is already taken into account in model (7.24)). For the gyroscope reported in [10], the position stiffness was measured to be 3.35×10^{-6} N m rad^{-1}. For the gyroscope reported in [8], the position stiffness is estimated below in Sect. 7.5, in which gyroscope resolution is studied. For the gyroscope reported in [12], it is assumed that the value of stiffness is the same as it was calculated for [8].

Analysis of Fig. 7.4 shows that the negative position stiffness dramatically increases a level of ARW by several orders of magnitude in compared to the same designs of gyroscopes having positive dynamic stiffness as shown in Fig. 7.3 and corresponds to the tactical grade performance. However, the variation of a value of stiffness itself does not lead to a significant change in AWR. Indeed, for a particular speed of rotation of the gyroscope [10], for instance, at 1000 rpm, changing the stiffness within a range of 3.35×10^{-7}–3.35×10^{-5} N m rad^{-1} corresponds to a change of ARW within a rang of 0.1–0.3° h$^{-1/2}$ It means that ARW is increased by a factor of 3, while the stiffness is increased by two orders of magnitude. Even, for the similar design reported [8], ARW has an alike level of ARW, although the stiffness for calculation is two orders of magnitude higher. The estimation of ARW for particular speeds of rotation is presented in Table 7.2. Analysis of Table 7.2 shows that mechanical thermal noise in gyroscopes reported in [8, 10] has an alike level and it is one order of magnitude less than measured one. However, for gyroscope reported in [12] for the same stiffness as it is in [8], a level of mechanical thermal noise becomes close to the measured value. It can be explained by a decrease of axial moment of inertia J_z by two orders of magnitude compared to [8], while the speed was increased by a factor of 7 only.

Another important issue, which comes from the analysis of (7.25) and Fig. 7.4, is that at a particular speed of rotation, which is defined by the transient speed:

Table 7.2 Angular random walk in electrostatic gyroscopes

References	Rotor speed, rpm	Stiffness, $\mathrm{N\,m\,rad^{-1}}$	Calculated ARW by (7.24)	Measured ARW
[10]	10085	3.35×10^{-6}	$0.014°\,h^{-1/2}$	$0.9°\,h^{-1/2}$
[8]	12300	4.0×10^{-4}	$0.02°\,h^{-1/2}$	$0.15°\,h^{-1/2}$
[12]	74000	4.0×10^{-4}	$0.1°\,h^{-1/2}$	$0.12°\,h^{-1/2}$

$$\Omega_t = \sqrt{\frac{c_s}{\kappa J}}, \tag{7.26}$$

the behaviour of ARW curve is transformed. This transformation occurs due to dynamical compensation of negative stiffness by means of positive stiffness because of rotation. This transient speed, Ω_t, leads to a new dynamical structure in gyroscope model (7.10) without dynamical stiffness, $C_d = 0$, and as a result a gyroscope has a new dynamical behaviour in whole. For instance, for the gyroscope reported in [10] and stiffness of $3.35 \times 10^{-6}\,\mathrm{N\,m\,rad^{-1}}$ the transient speed is to be 20642 rpm. This fact can provide an explanation of "motor-induced noise" effect (the name is given by the authors of article [11]) observed at 20000 rpm in [11].

7.4 Johnson Noise. Resolution

By considering the gyroscope scheme shown in Fig. 7.2, a relationship between the torque and angular speed of the rotor displacement can be written as

$$\tau = J\frac{d\dot{\alpha}}{dt} = j\omega J\dot{\alpha}, \tag{7.27}$$

both vectors being directed along the same radial axis lying on the equatorial surface of the rotor. Taking into account that a gyroscopic torque generated by the noise component of the measuring rate is $\tau = J_z\Omega \cdot \langle \Omega_N \rangle$, and equating its right-hand side to the right-hand term of Eq. (7.27), the noise component can be written in terms of the mean absolute square values as follows:

$$\langle \Omega_N^2 \rangle = \frac{2\omega_0^2 J^2}{\Omega^2 J_z^2}\langle \dot{\alpha}^2 \rangle. \tag{7.28}$$

In Eq. (7.28) the resonance frequency $\omega_0^2 = c_s/J$ is used for the same reason provided previously. It is worth noting that a fruitful discussion about a similar problem can be found in the Feynman Lectures [27, p. 41-2] and Post's article [28]. Knowing the relationship between the mean square of $\langle \dot{\alpha}^2 \rangle$ and the spectral density S_τ, defined by Eq. (7.12) through the square absolute value of the gyroscope impedance defined by Eq. (7.10), the Johnson noise for an ideal gyroscope over the entire frequency range can be written as

$$
\langle \Omega_{NJ}^2 \rangle = \frac{4\omega_0^2 \, J k_B T \mu \tilde{Q}}{\pi \Omega^2 J_z^2 \tilde{\mu} \tilde{\Omega}_0} \times
$$
$$
\int_{-\infty}^{\infty} \frac{d\omega}{\frac{\tilde{Q}^2}{\tilde{\Omega}_0^2} \left(\omega - \Omega - \frac{\tilde{\Omega}_0^2}{\omega - \Omega} \right)^2 + 2\frac{\tilde{Q}h}{\tilde{\Omega}_0 \tilde{\mu}} \left(\omega - \Omega - \frac{\tilde{\Omega}_0^2}{\omega - \Omega} \right) + 1}, \tag{7.29}
$$

where $\tilde{\Omega}_0^2 = C_d/J$, $\tilde{\mu}^2 = \mu^2 + h^2$ is the effective damping coefficient and $\tilde{Q} = J\tilde{\Omega}_0/\tilde{\mu}$ is the effective quality factor. To provide a practical viewpoint, the case of $\mu > 0$ is considered in Appendix C.2. The noise component of the measuring rate due to the Johnson noise is now

$$
\langle \Omega_{NJ} \rangle = \frac{\sqrt{4 k_B T \omega_0^2 J}}{J_z \Omega} \times \frac{180 \cdot 3600}{\pi}, \; {}^\circ/h. \tag{7.30}
$$

Equation (7.30) defines the resolution of an ideal two-axis gyroscope, and provides the limiting value for the minimum measuring rate [29, p. 16].

A similar result to (7.30) can be obtained by applying the Ising criterion [17], [19, p. 185]. G. Ising established that the energy of measurement must be greater than $4k_B T$ for a correct measurement. Considering the gyroscope under steady-state conditions means that some virtual experiment is carried out, in which the gyroscopic torque due to the measuring rate is changing slowly enough so that the following equation can be written

$$
(c_s - j\mu\Omega)\bar{\alpha} = -\bar{\tau}^*. \tag{7.31}
$$

According to the Ising criterion, the mean energy, which is accumulated by the system, must be

$$
\frac{1}{2} c_s \langle \alpha^2 \rangle \geq 4 k_B T. \tag{7.32}
$$

By multiplying (7.31) with its complex conjugate, the effect of position stiffness can be studied. Assume that c_s is much lager than $\mu\Omega$. Hence, the equation in terms of mean square values can be written as

$$
2c_s^2 \langle \alpha^2 \rangle = \langle \tau^2 \rangle. \tag{7.33}
$$

Accounting for (7.32), the resolution of the ideal gyroscope becomes

Table 7.3 Gyroscope resolution for design reported in [10]

Measured in [10]	Due to Johnson noise calculated by (7.30)	Ising criterion calculated by (7.34)
50o/h (0.014o/sec)	32o/h	65o/h

$$\langle \Omega_{NI} \rangle = \frac{\sqrt{16 k_B T \omega_0^2 J}}{J_z \Omega} \times \frac{180 \cdot 3600}{\pi}, \; ^\circ/\text{h}. \tag{7.34}$$

Comparing the resolution equations for a gyroscope based on Johnson noise (7.30) and the Ising criterion (7.32), it is seen that for both equations the main parameter, which restricts the resolution, is the stiffness, c_s, but the source of this noise is still the same thermal fluctuating torque. The difference in equations is only due to the coefficients, which are 2 and 4, respectively. Thus,

$$\langle \Omega_{NI} \rangle > \langle \Omega_{NJ} \rangle. \tag{7.35}$$

The resolution defined through Johnson noise reflects the fact that, if the damping coefficient is zero, then the noise disappears (please see Appendix C.2). The resolution based on the Ising criterion gives a qualitative estimation and it is independent from the value of the damping coefficient. It is worth noting that both mechanisms are independent of the dynamic stiffness and the sign of angular position stiffness.

7.5   Analysis of Resolution of Reported Gyroscopes

For all known prototypes reported in the literature, the angular position stiffness is preserved. For instance, for inductive micromachined contactless suspensions, the position stiffness lies in the range of 8.0×10^{-11} N m rad^{-1} [30] to 1.5×10^{-8} N m rad^{-1} [31], whereas electrostatic contactless suspensions cover a range from 1.0×10^{-6} N m rad^{-1} [10] to 1.5×10^{-5} N m rad^{-1}[32]. Let us estimate the resolution of the electrostatic gyroscope reported in [10]. The parameters of the gyroscope used for the estimation are: the radial moment of inertia, J, is 7.2×10^{-13} kg m^2, the angular momentum of the rotor, $J_z \Omega$, is 1.51×10^{-9} kg m^{2}s^{-1} and the measured position stiffness, c_s, is 3.35×10^{-6} N m rad^{-1}. The results are listed in Table 7.3, and are in a good agreement with the measurements.

For the same design of gyroscope, the effect of rotor speed on its resolution was studied in [11]. The results of the measurements and modelling are presented in Table 7.4. In this evaluation, it is assumed that the position stiffness shows only small variations with speed, hence it was treated as a constant. The analysis of the results shows that modelling agrees well with experimental data. In Fig. 7.5 the results of both Tables 7.3 and 7.4 are summarized and joined with theoretical curves. Note that

Table 7.4 Effect of rotor speed on gyroscope resolution

Rotor speed, rpm	Measured in [11]	Due to Johnson noise calculated by (7.30)	Ising criterion calculated by (7.34)
10000	65°/h 0.018°/sec	32°/h	65°/h
15000	43°/h (0.012°/sec)	22°/h	43°/h
20000	43°/h (0.012°/sec)	16°/h	32°/h

Table 7.5 Levitated two-axis gyroscopes

References	Type	Rotor speed, rpm	Angular momentum of rotor, kg m^{2}s^{-1}
[2]	Inductive	1000	6.72×10^{-13}
[4]		3035	3.95×10^{-11}
[8]	Electrostatic	12300	5.34×10^{-9}
[10]		10085	1.51×10^{-9}

Fig. 7.5 Resolution of the electrostatic gyroscope measured in [10, 11] together with theoretical curves

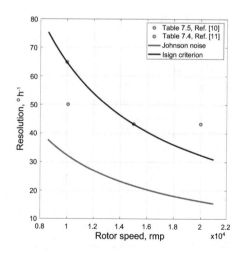

the measured resolution corresponded to a rotor speed of 2.0×10^4 rpm, which is larger than the modelling value, and has the same value as for the speed of 1.5×10^4. The authors in [11] explained this fact by "motor-induced noise". However, according to the developed theory, a decrease in resolution can be explained by an increase in the angular position stiffness just by a factor of 1.5 corresponding to 5.0×10^{-6} N m rad^{-1}.

Let us conduct the analysis of the effect of the position stiffness on the resolution of different levitated two-axis gyroscopes. For this purpose, the gyroscopes reported

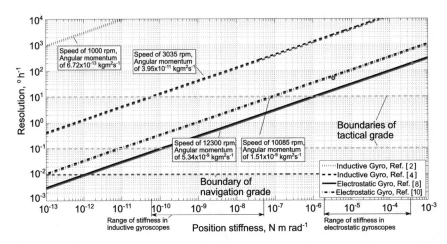

Fig. 7.6 Dependence of resolution in levitated two-axis gyroscopes on the value of the position stiffness: red and black lines correspond to a Johnson noise model (7.30) and an Ising criterion (7.34), respectively; the blue circle corresponds to resolution of $50°$/h measured for position stiffness of 3.35×10^{-6} N m rad^{-1} [10]; the green dot denotes a resolution of $180°$/h as reported in [8]

in [2, 4, 8, 10] are considered. Their design particularities are listed in Table 7.5. The result of the study is presented as a logarithmic plot of the position stiffness changing within the range of 1×10^{-13}–1×10^{-3} N m rad^{-1}, and a corresponding resolution which is varying over the range of 1×10^{-3}–$1 \times 10^{3}°$/h, covering rate, tactical and navigation grades, displayed in Fig. 7.6. Figure 7.6 demonstrates the dependence of the resolution of these gyroscopes on the position stiffness. The analysis shows that for a resolution corresponding to tactical grade, electrostatic gyroscopes must have a position stiffness of less than 1×10^{-6} N m rad^{-1} for a rotor speed of more than 12000 rpm, which is actually not so far from what has been achieved to date. However, a really challenging issue arises for navigation grade. For the considered electrostatic gyroscopes, the position stiffness must already be less than 1×10^{-12} N m rad^{-1}, which is six orders of magnitude less than that needed for tactical grades.

The position stiffness measured in inductive gyroscopes studied in [2] was 8×10^{-11} N m rad^{-1}. Unfortunately, the position stiffness in gyroscope [4] was not measured, but it can be calculated by using the analytical model developed in [31]. Thus, the calculated position stiffness of gyroscope reported in [4] should be around 7.41×10^{-8} N m rad^{-1}. Although the reported range of position stiffness values of inductive gyroscopes is several orders less than for electrostatic gyroscopes, even under normal operating temperatures, the resolution corresponds only to rate-grade performance.

In addition, an "inverse" problem can be solved by using the proposed theory. Upon knowing a measured value of gyroscope resolution, its position stiffness can be estimated. For instance, the measured resolution of the gyroscope reported in [8] was $180°$/h. Hence, the position stiffness of the gyroscope [8] should be around 4.0×10^{-4} N m rad^{-1} as shown in Fig. 7.6.

Fig. 7.7 Comparison of the scale factors of levitated and vibratory gyroscopes: sense-mode quality factors Q are also plotted for the vibratory gyroscopes [33, 34]

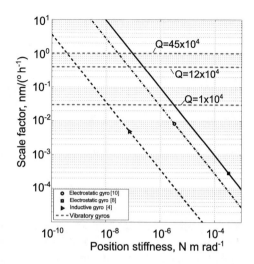

7.6 Scale Factor

The scale factor of the open-loop levitated gyroscope relative to its one measuring axis is [16]

$$SF = \frac{J_z \Omega}{c_s}, \ sec, \tag{7.36}$$

which is also dependent on the position stiffness. There is an interest to study it separately in the context of resolution. Also, we can provide a comparison with the scale factor of Coriolis vibratory gyroscopes, which have already reached tactical grade [35]. In order to do so, the scale factor of levitated gyroscope can be represented as follows:

$$SF = 2\frac{J_z \Omega}{c_s} \frac{2r - w}{2} \frac{\pi}{180 \cdot 3600} \cdot 10^9, \ nm/(°/h), \tag{7.37}$$

where r is the radius of the rotor, and w is the width of the electrodes of a capacitive linear displacement sensor. It is assumed that the measurements of displacement occur near to an outer edge of an axially symmetric rotor. Now, using (7.37), the direct comparison of levitated gyroscope SF with vibratory gyroscopes [33, 34] can be performed.

Table 7.6 lists the parameters of the levitated gyroscopes and the results of the scale factor calculations. Figure 7.7 shows the dependence of the scale factor on the angular position stiffness, together with the level of SF measured in vibratory gyroscopes for different quality factors [33, 34]. Analysis of Table 7.6 reveals that the levitated gyroscope reported in [10] has a maximum scale factor of $8.3 \times 10^{-3} \ nm/(°/h)$ compared with the gyroscopes [4, 8]. However, the maximum scale factor of the levitated gyroscope is one order of magnitude less than for the single mass vibratory gyroscope with quality factor of 1×10^4, and three orders of magnitude less than

Table 7.6 Scale factor of levitated gyroscopes

Design parameters	Electrostatic		Inductive
	Ref. [10]	Ref. [8]	Ref. [4]
Radius, r, m	2×10^{-3}	2×10^{-3}	1.1×10^{-3}
Width, w, μm	215	300	300
Rotor Speed, rmp	10085	12300	3035
Stiffness, c_s, N m rad^{-1}	3.35×10^{-6}	4.0×10^{-4}	7.41×10^{-8}
Scale Factor, $nm/(°/h)$	8.6×10^{-3}	2.9×10^{-4}	5.2×10^{-3}

for the quadruple mass vibratory gyroscope with quality factor of 45×10^4. In fact, the vibratory gyroscope with an *SF* of more than 1 nm/(°/h) successfully demonstrated tactical grade performance [35]. Thus, the scale factor provides an additional requirement for the value of the position stiffness. For instance, in the framework of designs of electrostatic gyroscopes [10], the position stiffness must be less than 1.0×10^{-8} N m rad^{-1}, in order to satisfy requirements of tactical grade for both resolution as well as scale factor.

7.7 Mechanical Thermal Noise in Vibratory and Levitated Gyroscopes

It is seen from above that the scale factor of both types of gyroscopes can be expressed through the same dimension, in particular, written into nm/(°/h). This dimension was proposed by *A. Trusov et al.* [34] and becomes the appropriate one for the applied analysis. Then, rewriting equations for mechanical thermal noise in terms of *SF* having the mentioned above dimension, we have an opportunity for a direct comparison of performances in both types of gyroscopes through the scale factor. This fact provides a convenient way in order to show a general view of the influence of MTN on performances in reported prototypes of micro-gyroscopes, and potential accuracies of future ones, as well as a dependence of a level of MTN in a micro-gyroscope on its design.

Inasmuch as the scale factor in levitated gyroscopes can be controlled by the position stiffness and angular momentum of a rotor, hence in order to represent MTN in such gyroscopes through the scale factor, two manners can be recognized. The first manner corresponds to controlling the position stiffness and keeping a constant angular momentum of the rotor. Hence, the Johnson noise in levitated gyroscopes (7.30) can be rewritten as follows:

$$\langle \Omega_{NJ} \rangle_I = \sqrt{\frac{4k_B T (2r - w)}{SF J_z \Omega}} \times \sqrt{\frac{180 \cdot 3600 \cdot 10^9}{\pi}}, \; °/h. \tag{7.38}$$

The second manner is vice versa to the previous one. Thus, we have

$$\langle \Omega_{NJ} \rangle_{II} = \frac{(2r - w)}{SF} \sqrt{\frac{4k_B T}{c_s}} \times 10^9, \; °/h, \tag{7.39}$$

where SF (in both equations) is defined by (7.37) and has the dimension of nm/(°/h). Mechanical thermal noise in vibratory gyroscopes in terms of SF written into nm/(°/h) is described by [34]

$$\langle \Omega_N \rangle_v = \frac{1}{SF} \sqrt{\frac{k_B T}{\omega_y^2 M4}} \times 1.455 \cdot 10^9, \; °/h, \tag{7.40}$$

where M is the mass of structure in kg, ω_y is the mode-matched operational frequency in rad/s.

Finally, the comparative analysis of MTN in micro-gyroscopes is reduced to the analysis of the scale factor in both types of gyroscopes and the sources of its constrain. For the further discussion, the results of theoretical and experimental study of family of ultra-high Q-factor silicon MEMS vibratory gyroscopes developed in the Shkel group are used [33–37]. We compare these results with MTN arising in a levitated gyroscope based on the design reported by *F. Han et al.* [10]. MTN in both types of micro-gyroscopes as a function of the scale factor corresponding to Eq. (7.38), (7.39) and (7.40) is shown in Fig. 7.8.

In general, the analysis of Fig. 7.8 shows that independent of the gyroscope type in order to demonstrate navigation-grade performance in a micro-gyroscope the scale factor must be larger than 10 nm/(°/h). The scale factor close to this value was obtained in vibratory gyroscopes. In particular the quad mass gyroscope with Q-factor better than 1.7 million and the measured bias stability of 0.09(°/h) was reported in [36]. The scale factor in this gyroscope is about 5 nm/(°/h). The recent result, where the vibratory gyroscope with Q-factor over 2 million [37] corresponding to the largest value of the scale factor about 6 nm/(°/h) in compared to reported micro-gyroscopes in the literature, establishes the physical limit, to further increase the scale factor, due to the fundamental thermoelastic damping [36]. It is an obvious fact that a vibratory gyroscope with the scale factor higher than 10 nm/(°/h) becomes a challenge requiring a resonator with smaller dissipation energy.

Although, the scale factor in reported prototypes of levitated gyroscopes is several orders of magnitude less than in vibratory gyroscopes, however, controlling position stiffness and angular momentum of the rotor in the framework of the existed gyroscope design the scale factor can reach 10 nm/(°/h), which is limited by the value J_z/μ. But the maximum scale factor can be still increased by vacuum. As it is seen

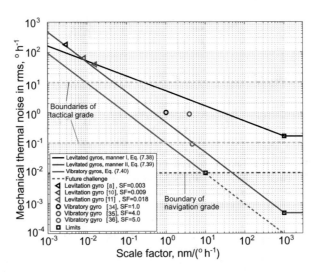

Fig. 7.8 Comparison of mechanical thermal noise in vibratory and levitated gyroscopes given in the root mean square (rms) value versus the scale factor

from the analysis of Fig. 7.8, the potential performance in existed prototypes of levitated gyroscopes can be beyond the navigation grade.

References

1. C. Shearwood, C. Williams, P. Mellor, R. Yates, M. Gibbs, A. Mattingley, Levitation of a micromachined rotor for application in a rotating gyroscope. Electron. Lett. **31**(21), 1845–1846 (1995)
2. C. Shearwood, K.Y. Ho, C.B. Williams, H. Gong, Development of a levitated micromotor for application as a gyroscope. Sensor. Actuat. A-Phys. **83**(1–3), 85–92 (2000)
3. X. Wu, W. Chen, X. Zhao, W. Zhang, Micromotor with electromagnetically levitated rotor using separated coils. Electron. Lett. **40**(16), 996–997 (2004)
4. W. Zhang, W. Chen, X. Zhao, X. Wu, W. Liu, X. Huang, S. Shao, The study of an electromagnetic levitating micromotor for application in a rotating gyroscope. Sensor. Actuat. A-Phys. **132**(2), 651–657 (2006)
5. W. Liu, W.-Y. Chen, W.-P. Zhang, X.-G. Huang, Z.-R. Zhang, Variable-capacitance micromotor with levitated diamagnetic rotor. Electron. Lett. **44**(11), 681–683 (2008)
6. Y. Su, Z. Xiao, C.B. Williams, K. Takahata, Micromachined graphite rotor based on diamagnetic levitation. IEEE Electron Device Lett. **36**(4), 393–395 (2015). https://doi.org/10.1109/LED.2015.2399493
7. G. He, K. Chen, S. Tan, W. Wang, Electrical levitation for micromotors, microgyroscopes and microaccelerometers. Sens. Actuators A: Phys. **54**(1), 741–745 (1996)
8. T. Murakoshi, Y. Endo, K. Fukatsu, S. Nakamura, M. Esashi, Electrostatically levitated ring-shaped rotational gyro/accelerometer. Jpn. J. Appl. Phys. **42**(4B), 2468–2472 (2003)
9. B. Damrongsak, M. Kraft, S. Rajgopal, M. Mehregany, Design and fabrication of a micromachined electrostatically suspended gyroscope. Proc. Inst. Mech. Eng. Part C: J. Mech. Eng. Sci. **222**(1), 53–63 (2008)

10. F. Han, Y. Liu, L. Wang, G. Ma, Micromachined electrostatically suspended gyroscope with a spinning ring-shaped rotor. J. Micromechanics Microengineering **22**(10), 105032 (2012)

11. B. Sun, F. Han, L. Li, Q. Wu, Rotation control and characterization of high-speed variable-capacitance micromotor supported on electrostatic bearing. IEEE Trans. Ind. Electron. **63**(7), 4336–4345 (2016). https://doi.org/10.1109/TIE.2016.2544252

12. S. Nakamura, MEMS inertial sensor toward higher accuracy & multi-axis sensing, in *Proc. 4th IEEE Conf. on Sensors* (Irvine, CA 2005), pp. 939–942. https://doi.org/10.1109/ICSENS. 2005.1597855

13. K. Liu, W. Zhang, W. Liu, W. Chen, K. Li, F. Cui, S. Li, An innovative micro-diamagnetic levitation system with coils applied in micro-gyroscope. Microsyst. Technol. **16**(3), 431–439 (2010)

14. Y. Xu, Q. Cui, R. Kan, H. Bleuler, J. Zhou, Realization of a diamagnetically levitating rotor driven by electrostatic field. IEEE/ASME Trans. Mechatron. **22**(5), 2387–2391 (2017). https:// doi.org/10.1109/TMECH.2017.2718102

15. V. Annovazzi-Lodi, S. Merlo, Mechanical-thermal noise in micromachined gyros. Microelectron. J. **30**(12), 1227–1230 (1999)

16. K. Poletkin, U. Wallrabe, Static behavior of closed-loop micromachined levitated two-axis rate gyroscope. IEEE Sensors J. **15**(12), 7001–7008 (2015). https://doi.org/10.1109/JSEN. 2015.2469715

17. G. Ising, Les Fluctuation dans les Appareil de Mesues. Philos. Mag. Lond. **6**(51), 827 (1926)

18. K.V. Poletkin, A.I. Chernomorsky, C. Shearwood, A proposal for micromachined dynamically tuned gyroscope, based on contactless suspension. IEEE Sens. J. **12**(06), 2164–2171 (2012). https://doi.org/10.1109/JSEN.2011.2178020

19. L. Brillouin, *Science and Information Theory*, 2nd edn. (Academic Press Inc., New York, 1962)

20. D. Ormandy, L. Maunder, Dynamics of oscillogyro. J. Mech. Eng. Sci. **15**(3), 210–217 (1973)

21. L.I. Brozgul, E.L. Smirnov, The history of the mechanics of gyroscopic systems, in (Izdatel'stvo Nauka, 1975) Chap. Vibrational gyroscopes, pp. 43–60

22. L. Maunder, Dynamically tuned gyroscopes, in *5th World Congress on Theory of Machines and Mechanisms* (New York, 1979), pp. 470–473

23. H. Nyquist, Thermal agitation of electric charge in conductors. Phys. Rev. **32**(1), 110–113 (1928)

24. R. Leland, Mechanical-thermal noise in MEMS gyroscopes. IEEE Sens. J. **5**(3), 493–500 (2005)

25. K.V. Poletkin, A.I. Chernomorsky, C. Shearwood, Selection of series compensation parameters for closed loop rotor vibratory gyroscope. IEEE Sens. J. **12**(05), 1384–1392 (2012). https:// doi.org/10.1109/JSEN.2011.2169784

26. A. Kazamarov, A. Palatnik, L. Rodnyanskiy, *Dynamics of Two Axis Automatic Control System* (Gos Izd-vo Fiz-Mat. Lit-ry, Moscow, 1967 (in Russian))

27. R.P. Feynman, R.B. Leighton, M. Sands, *The Feynman Lectures on Physics*, vol. I (Addison-Wesley Publishing Company Inc, Reading, Massachusetts, 1963)

28. E.R. Post, G.A. Popescu, N. Gershenfeld, Inertial measurement with trapped particles: a micro-dynamical system. Appl. Phys. Lett. **96**(14), 143501 (2010). https://doi.org/10.1063/1.3360808

29. IEEE Standard for Inertial Sensor Terminology, IEEE Std 528-2001, 0–1– (2001). https://doi. org/10.1109/IEEESTD.2001.93360

30. C.B. Williams, C. Shearwood, P.H. Mellor, Modeling and testing of a frictionless levitated micromotor. Sensor. Actuat. A-Phys. **61**, 469–473 (1997)

31. Z. Lu, K. Poletkin, B. den Hartogh, U. Wallrabe, V. Badilita, 3D micro-machined inductive contactless suspension: testing and modeling. Sens. Actuators A: Phys. **220**, 134–143 (2014). https://doi.org/10.1016/j.sna.2014.09.017

32. F. Han, L. Wang, Q. Wu, Y. Liu, Performance of an active electric bearing for rotary micromotors. J. Micromech. Microeng. **21**, 085027 (2011)

33. A.A. Trusov, I.P. Prikhodko, S.A. Zotov, A.R. Schofield, A.M. Shkel, Ultra-high Q silicon gyroscopes with interchangeable rate and whole angle modes of operation, in *SENSORS, 2010 IEEE* (2010), pp. 864–867

34. A.A. Trusov, A.R. Schofield, A.M. Shkel, Micromachined rate gyroscope architecture with ultra-high quality factor and improved mode ordering. Sens. Actuators A: Phys. **165**(1), 26–34 (2011)
35. I.P. Prikhodko, S. Zotov, A. Trusov, A. Shkel, Sub-degree-per-hour silicon MEMS rate sensor with 1 million Q-factor, in, 16th International Solid-State Sensors. Actuators and Microsystems Conference **2011**, 2809–2812 (2011)
36. S. Askari, M. Asadian, K. Kakavand, A. Shkel, Near-navigation grade quad mass gyroscope with Q-factor limited by thermo-elastic damping. Energy **44**, 1–125 (2016)
37. M.H. Asadian, S. Askari, A.M. Shkel, An ultra-high vacuum packaging process demonstrating over 2 million Q-factor in MEMS vibratory gyroscopes, in *IEEE Sensors Letters* (2017)

Appendix A
Mathematical Notation

Vectors are denoted by bold letters. Using the index notation, in a cartesian reference frame with a standard basis having unit vectors e_1, e_2 and e_3, a *vector a* is represented through a linear combination of unit vectors as

$$a = a_1 e_1 + a_2 e_2 + a_3 e_3, \tag{A.1}$$

where a_1, a_2 and a_3 are scalar quantities introducing *elements* of a. Three unit vectors generate *cartesian reference basis*, which is designated by a symbol $\underline{e}$. A line under the letter e stands for a matrix, elements of which are vectors. In a particular case, we have a column-matrix or $\underline{e} = [e_1 \ e_2 \ e_3]^T$. The exponent ^{T} denotes *transposition*. The basis vectors are mutually orthogonal and fulfil the conditions

$$e_i \cdot e_j = \delta_{ij} \quad (i, j = 1, 2, 3), \tag{A.2}$$

where δ_{ij} is the *Kronecker's delta* and a symbol of " $\cdot$ " is the *scalar product* of two unit vectors.

In this work, we use more than one vector basis. To distinguish one standard vector basis from another one, a right superscript, which can be designated by either a letter or a number, is used, for instance, $\underline{e}^1$, $\underline{e}^2$, $\underline{e}^x$ and $\underline{e}^y$ and so on. Suppose that e_i^p ($i = 1, 2, 3$) are the unit vectors of the vector basis $\underline{e}^p$, while e_i^q ($i = 1, 2, 3$) are the unit vectors of the vector basis $\underline{e}^q$. Then the basis $\underline{e}^p$ can be represented as a linear combination of the basis $\underline{e}^q$ as follows:

$$e_i^p = \sum_{j=1}^{3} a_{ij}^{pq} e_j^q \quad (i = 1, 2, 3), \tag{A.3}$$

where the scalar values a_{ij}^{pq} are the *direction cosines* of three basis vectors:

$$a_{ij}^{pq} = e_i^p \cdot e_j^q = \cos \sphericalangle \left(e_i^p, e_j^q \right) \quad (i, j = 1, 2, 3). \tag{A.4}$$

© The Editor(s) (if applicable) and The Author(s), under exclusive license
to Springer Nature Switzerland AG 2021
K. Poletkin, *Levitation Micro-Systems*, Microsystems and Nanosystems,
https://doi.org/10.1007/978-3-030-58908-0

Equation (A.3) can be written as a matrix equation:

$$\underline{e}^p = \underline{\Lambda}^{pq}\underline{e}^q,$$ (A.5)

where $\underline{\Lambda}^{pq}$ is the (3×3) matrix of direction cosines. A line under letter Λ means a matrix. In general, a matrix can be thought as a table of real or complex numbers, or vectors. Note that the subscription of the matrix consists of two letters, namely, p and q, corresponding to the notations of the vector bases of $\underline{e}^p$ and $\underline{e}^q$, respectively. Their sequence in the subscript of the matrix shows that the matrix transforms the basis $\underline{e}^q$ into $\underline{e}^p$. If elements of a matrix are vectors, then for such matrix, both the *dot product* and the *cross product* are applicable. For instance, $\underline{e}^p \cdot \underline{e}^{pT} = \underline{E}$, where $\underline{E}$ is the unit matrix. Using the result of this example, we apply the dot product of $\underline{e}^{pT}$ to both sides of Eq. (A.5). Then, the equation for the direction cosine matrix is obtained in an explicit form such as

$$\underline{\Lambda}^{pq} = \underline{e}^p \cdot \underline{e}^{qT}.$$ (A.6)

Let us show that the matrix of direction cosines is orthogonal. Indeed, using the orthogonality condition (A.2) for vector bases, we have

$$\delta_{ij} = e_i^p \cdot e_j^p = \left(\sum_{n=1}^{3} a_{in}^{pq} e_n^q\right) \cdot \left(\sum_{m=1}^{3} a_{jm}^{pq} e_m^q\right) = \sum_{n=1}^{3}\sum_{m=1}^{3} a_{in}^{pq} a_{jm}^{pq} \left(e_n^q \cdot e_m^q\right)$$
$$= \sum_{n=1}^{3} a_{in}^{pq} a_{jn}^{pq} \quad (i, j = 1, 2, 3),$$ (A.7)

or in a matrix form:

$$\underline{\Lambda}^{pq}\underline{\Lambda}^{pqT} = \underline{E}.$$ (A.8)

The last equation shows the important property of orthogonal matrices that its transpose matrix is also an *inverse matrix*:

$$\left(\underline{\Lambda}^{pq}\right)^{-1} = \underline{\Lambda}^{pqT}.$$ (A.9)

Let us again consider the vector $\boldsymbol{a}$ and present its elements as the column-matrix $\underline{a} = [a_1 \ a_2 \ a_3]^T$. Then the vector can be written in a matrix form as follows:

$$\boldsymbol{a} = \underline{e}^T\underline{a},$$ (A.10)

or

$$\boldsymbol{a} = \underline{a}^T\underline{e}.$$ (A.11)

Elements of vector $\boldsymbol{a}$ in two different bases, for instance, in basis $\underline{e}^p$ and $\underline{e}^q$, have different values and denoted by $\underline{a}^p$ and $\underline{a}^q$, respectively. With reference on (A.10), the following equation is true:

$$a = \underline{e}^{p^T} \underline{a}^p = \underline{e}^{q^T} \underline{a}^q. \tag{A.12}$$

Applying the dot product of $\underline{e}^p$ from the left side to both last terms of the equation above and accounting for (A.6), we can write

$$\underline{a}^p = \underline{\Lambda}^{pq} \underline{a}^q. \tag{A.13}$$

This equation states the rule for the transformation of vector elements.

In general, a vector can have n elements defined on the standard basis $\underline{e}$ with n-number of unit vectors. Then a matrix of transformation for such a vector has the size $n \times n$. Since a number of rows and columns of the matrix is the same number n, then the matrix is called *square* one and the number n of its rows (columns) is called *the order of the matrix*. In applications of square matrices, the mathematical apparatus of *determinants* is widely used.

The *determinant* of the square matrix of n-th order is called the algebraic sum of products of matrix elements, each of them is taken from one row, one column, and aside each product has a positive or negative sign, which is defined by a certain rule. The determinant of the square matrix $\underline{A}$ is denoted by $\det \underline{A}$ or $|\underline{A}|$. Then, the definition of determinant given above can be written as the following equation:

$$\det \underline{A} = \begin{vmatrix} a_{11} & a_{12} & \dots & a_{1n} \\ a_{21} & a_{21} & \dots & a_{2n} \\ \vdots & \vdots & \ddots & \vdots \\ a_{n1} & a_{n1} & \dots & a_{nn} \end{vmatrix} = \sum_{\sigma} \operatorname{sgn} \sigma \prod_{i=1}^{n} a_{1\sigma(i)}, \tag{A.14}$$

where the sum runs over all $n!$ permutations σ of the n terms and sgn σ is the sign of permutations σ equaling to $+1$ if the transposition is even or -1 if it is odd.

Also, the determinant can be defined inductively through the *Laplace expansion* in the following way. Let $\underline{A}_{ij}$ denote the sub-matrix of $\underline{A}$, which is formed by deleting row i and column j in the matrix $\underline{A}$. Hence,

$$\det \underline{A} = \sum_{j=1}^{n} (-1)^{i+j} a_{ij} \det \underline{A}_{ij} = \sum_{i=1}^{n} (-1)^{i+j} a_{ij} \det \underline{A}_{ij}, \tag{A.15}$$

where $i \leq n$ and $j \leq n$; $\det \underline{A}_{ij}$ is so-called the minor of the order of $n - 1$. Now using the Laplace expansion, we can inductively present determinants of square matrices beginning with a 1-by-1 matrix. Hence,

$$|a_{11}| = a_{11}$$

$$\begin{vmatrix} a_{11} & a_{12} \\ a_{21} & a_{21} \end{vmatrix} = a_{11}a_{22} - a_{12}a_{21}$$

$$\begin{vmatrix} a_{11} & a_{12} & a_{13} \\ a_{21} & a_{22} & a_{23} \\ a_{31} & a_{32} & a_{33} \end{vmatrix} = \begin{aligned} a_{11}a_{22}a_{33} + a_{12}a_{23}a_{31} + a_{13}a_{21}a_{32} \\ -a_{11}a_{23}a_{32} - a_{12}a_{21}a_{33} - a_{13}a_{22}a_{31} \end{aligned} \qquad (A.16)$$

and so on. Below, we discuss some properties of the determinants.

1. The determinant of the transposed matrix is equal to the determinant of the given matrix, that is,

$$\det \underline{A}^T = \det \underline{A}. \qquad (A.17)$$

This property follows directly from the Laplace expansion (A.15).

2. The determinant of the product of two matrices is equal to the product of the determinants of the individual matrices:

$$\det(\underline{A}\,\underline{B}) = \det \underline{A} \det \underline{B}. \qquad (A.18)$$

3. The determinant of an orthogonal matrix is equal to ± 1. For instance, according to (A.18) and (A.8), the determinant of the matrix of direction cosines $\underline{\Lambda}^{pq}$ is equal to 1:

$$\det \underline{\Lambda}^{pq} = +1. \qquad (A.19)$$

4. If the elements of a row are represented as a sum of m summands, then the determinant of such the matrix is equal to the sum of m determinants in each of them has the corresponding summands of the noted row. Mathematically, this property can be written as

$$\begin{vmatrix} a_{11} & a_{12} & \cdots & a_{1n} \\ a_{21} & a_{22} & \cdots & a_{2n} \\ \vdots & \vdots & \ddots & \vdots \\ \sum_{l=1}^{m} a_{i1l} & \sum_{l=1}^{m} a_{i2l} & \cdots & \sum_{l=1}^{m} a_{inl} \\ \vdots & \vdots & \ddots & \vdots \\ a_{n1} & a_{n2} & \cdots & a_{nn} \end{vmatrix} = \sum_{l=1}^{m} \begin{vmatrix} a_{11} & a_{12} & \cdots & a_{1n} \\ a_{21} & a_{22} & \cdots & a_{2n} \\ \vdots & \vdots & \ddots & \vdots \\ a_{i1l} & a_{i2l} & \cdots & a_{inl} \\ \vdots & \vdots & \ddots & \vdots \\ a_{n1} & a_{n2} & \cdots & a_{nn} \end{vmatrix}. \qquad (A.20)$$

Indeed, using the Laplace expansion (A.15), let us write the determinant of the matrix as follows:

$$\begin{vmatrix} a_{11} & a_{12} & \cdots & a_{1n} \\ a_{21} & a_{22} & \cdots & a_{2n} \\ \vdots & \vdots & \ddots & \vdots \\ \sum_{l=1}^{m} a_{i1l} & \sum_{l=1}^{m} a_{i2l} & \cdots & \sum_{l=1}^{m} a_{inl} \\ \vdots & \vdots & \ddots & \vdots \\ a_{n1} & a_{n2} & \cdots & a_{nn} \end{vmatrix} = \sum_{j=1}^{n} (-1)^{i+j} \left(\sum_{l=1}^{m} a_{ijl} \right) \det \underline{A}_{ij}. \quad (A.21)$$

The right-hand side of Eq. (A.21) can be rewritten as follows: $\sum_{l=1}^{m} \left(\sum_{j=1}^{n} (-1)^{i+j} a_{ijl} \det \underline{A}_{ij} \right)$. This fact proves the property 4.

5. If all elements of a row are multiplied by a scalar, then the scalar can be moved over the sign of the determinant.

$$\begin{vmatrix} a_{11} & a_{12} & \cdots & a_{1n} \\ a_{21} & a_{21} & \cdots & a_{2n} \\ \vdots & \vdots & \ddots & \vdots \\ Ca_{i1} & Ca_{i2} & \cdots & Ca_{in} \\ \vdots & \vdots & \ddots & \vdots \\ a_{n1} & a_{n1} & \cdots & a_{nn} \end{vmatrix} = C \begin{vmatrix} a_{11} & a_{12} & \cdots & a_{1n} \\ a_{21} & a_{21} & \cdots & a_{2n} \\ \vdots & \vdots & \ddots & \vdots \\ a_{i1} & a_{i2} & \cdots & a_{in} \\ \vdots & \vdots & \ddots & \vdots \\ a_{n1} & a_{n1} & \cdots & a_{nn} \end{vmatrix}. \quad (A.22)$$

6. The determinant of a square matrix having two identical rows is equal to zero. Indeed, let us assume that a determinant of a square matrix $\underline{A}$ has two identical rows denoted by indexes f and i. Hence, according to the statement, we have $a_{f1} = a_{i1}$, $a_{f2} = a_{i2}$, $a_{f3} = a_{i3}$, ... , $a_{fn} = a_{in}$. The index i is larger than f, or $i > f$. Let us apply the Laplace expansion to the determinant and write it through minors along row i:

$$\det \underline{A} = \begin{vmatrix} a_{11} & a_{12} & \cdots & a_{1n} \\ \vdots & \vdots & \ddots & \vdots \\ a_{f1} & a_{f1} & \cdots & a_{fn} \\ \vdots & \vdots & \ddots & \vdots \\ a_{i1} & a_{i2} & \cdots & a_{in} \\ \vdots & \vdots & \ddots & \vdots \\ a_{n1} & a_{n1} & \cdots & a_{nn} \end{vmatrix} = \sum_{j=1}^{n} (-1)^{i+j} a_{ij} \det \underline{A}_{ij}. \quad (A.23)$$

Each minor $\det \underline{A}_{ij}$ of the $(n-1)$ order can be further expanded through minors of the $(n-2)$ order along row f:

$$\det \underline{A}_{ij} = \begin{vmatrix} a_{11} & \cdots & a_{1,j-1} & a_{1,j+1} & \cdots & a_{1n} \\ \vdots & \ddots & \vdots & \vdots & \ddots & \vdots \\ a_{f1} & \cdots & a_{f,j-1} & a_{f,j+1} & \cdots & a_{fn} \\ \vdots & \ddots & \vdots & \vdots & \ddots & \vdots \\ a_{n1} & \cdots & a_{n,j-1} & a_{n,j+1} & \cdots & a_{nn} \end{vmatrix} = \sum_{l=1}^{j-1}(-1)^{f+l}a_{fl}\det \underline{A}_{fl}^{ij} + \sum_{l=j+1}^{n}(-1)^{f+l-1}a_{fl}\det \underline{A}_{fl}^{ij}.$$

(A.24)

In minor $\det \underline{A}_{fl}^{ij}$, the superscript ij is used to remind that in this minor, the i row and j column as well as the f row and l column are absented. A structure of the minor $\det \underline{A}_{fl}^{ij}$ is as follows:

$$\det \underline{A}_{fl}^{ij} = \begin{vmatrix} a_{11} & \cdots & a_{1,f-1} & a_{1,f+1} & \cdots & a_{1,j-1} & a_{1,j+1} & \cdots & a_{1n} \\ \vdots & \ddots & \vdots & \vdots & \ddots & \vdots & & \ddots & \vdots \\ a_{f-1,1} & \cdots & a_{f-1,f-1} & a_{f-1,f+1} & \cdots & a_{f-1,j-1} & a_{f-1,j+1} & \cdots & a_{f-1,n} \\ a_{f+1,1} & \cdots & a_{f+1,f-1} & a_{f+1,f+1} & \cdots & a_{f+1,j-1} & a_{f+1,j+1} & \cdots & a_{f+1,n} \\ \vdots & \ddots & \vdots & \vdots & \ddots & \vdots & & \ddots & \vdots \\ a_{n1} & \cdots & a_{n,f-1} & a_{n,f+1} & \cdots & a_{n,j-1} & a_{n,j+1} & \cdots & a_{nn} \end{vmatrix}.$$

(A.25)

Analysis of (A.25) shows that minor $\det \underline{A}_{fl}^{ij} = \det \underline{A}_{fj}^{il}$. Due to the statement and the fact that the indexes f and i are fixed. They can be hidden for further study. Also, minors $\det \underline{A}_{fl}^{ij}$ are denoted as Δjl. Accounting for (A.24) and new notations, the determinant can be calculated by

$$\det \underline{A} = \sum_{j=1}^{n}(-1)^{j}a_{j}\left[\sum_{l=1}^{j-1}(-1)^{l}a_{l}\Delta_{jl} + \sum_{l=j+1}^{n}(-1)^{l-1}a_{l}\Delta_{jl}\right].$$

(A.26)

Performing summation in (A.26), we can write

$$\det \underline{A} = (-1)^{1}a_{1}\left[(-1)^{2-1}a_{2}\Delta_{12} + (-1)^{3-1}a_{3}\Delta_{13} + \ldots + (-1)^{n-1}a_{n}\Delta_{1n}\right]$$

(A.27)

$$+(-1)^{2}a_{2}\left[(-1)^{1}a_{1}\Delta_{21} + (-1)^{3-1}a_{3}\Delta_{23} + \ldots + (-1)^{n-1}a_{n}\Delta_{2n}\right]$$

$$\ldots$$

$$+(-1)^{n-1}a_{n-1}\left[(-1)^{1}a_{1}\Delta_{n-1,1} + (-1)^{2}a_{2}\Delta_{n-1,2} + \ldots + (-1)^{n-1}a_{n}\Delta_{n-1,n}\right]$$

$$+(-1)^{n}a_{n}\left[(-1)^{1}a_{1}\Delta_{n-1,1} + (-1)^{2}a_{2}\Delta_{n-1,2} + \ldots + (-1)^{n-1}a_{n-1}\Delta_{n,n-1}\right].$$

Carrying out the multiplications and rearranging terms in (A.27), and taking into account the symmetry of minors $\Delta_{jl} = \Delta_{lj}$, Eq. (A.27) can be rewritten as follows:

$$\det \underline{A} = \left[(-1)^{1+2-1} + (-1)^{2+1}\right] a_1 a_2 \Delta_{12} + \left[(-1)^{1+3-1} + (-1)^{3+1}\right] a_3 a_1 \Delta_{13} + ... \text{(A.28)}$$
$$+ \left[(-1)^{1+n-1} + (-1)^{1+n}\right] a_1 a_n \Delta_{1n} + ... + \left[(-1)^{2+n-1} + (-1)^{n+2}\right] a_2 a_n \Delta_{2n}$$
$$... + \left[(-1)^{n-1+n-1} + (-1)^{n+n-1}\right] a_{n-1} a_n \Delta_{n-1,n} = 0.$$

All terms in brackets are zeros, thus the property is proved.

7. The determinant is equal to zero, if elements of one of its row is proportional to elements of another one. Suppose we have a matrix:

$$\begin{vmatrix} a_{11} & a_{12} & ... & a_{1n} \\ \vdots & \vdots & \ddots & \vdots \\ Ca_{i1} & Ca_{i2} & ... & Ca_{in} \\ \vdots & \vdots & \ddots & \vdots \\ a_{i1} & a_{i2} & ... & a_{in} \\ \vdots & \vdots & \ddots & \vdots \\ a_{n1} & a_{n2} & ... & a_{nn} \end{vmatrix}. \tag{A.29}$$

Using properties 4 and 6, we can write

$$C \begin{vmatrix} a_{11} & a_{12} & ... & a_{1n} \\ \vdots & \vdots & \ddots & \vdots \\ a_{i1} & a_{i2} & ... & a_{in} \\ \vdots & \vdots & \ddots & \vdots \\ a_{i1} & a_{i2} & ... & a_{in} \\ \vdots & \vdots & \ddots & \vdots \\ a_{n1} & a_{n2} & ... & a_{nn} \end{vmatrix} = 0. \tag{A.30}$$

8. The determinant does not change its value, if we add elements multiplied by a scalar of one row to another one. Indeed,

$$\begin{vmatrix} a_{11} & \cdots & a_{1n} \\ \vdots & \ddots & \vdots \\ a_{i1}+Ca_{k1} & \cdots & a_{kn}+Ca_{in} \\ \vdots & \ddots & \vdots \\ a_{k1} & \cdots & a_{kn} \\ \vdots & \ddots & \vdots \\ a_{n1} & \cdots & a_{nn} \end{vmatrix} = \begin{vmatrix} a_{11} & \cdots & a_{1n} \\ \vdots & \ddots & \vdots \\ a_{i1} & \cdots & a_{kn} \\ \vdots & \ddots & \vdots \\ a_{k1} & \cdots & a_{kn} \\ \vdots & \ddots & \vdots \\ a_{n1} & \cdots & a_{nn} \end{vmatrix} + C \begin{vmatrix} a_{11} & \cdots & a_{1n} \\ \vdots & \ddots & \vdots \\ a_{k1} & \cdots & a_{in} \\ \vdots & \ddots & \vdots \\ a_{k1} & \cdots & a_{kn} \\ \vdots & \ddots & \vdots \\ a_{n1} & \cdots & a_{nn} \end{vmatrix} = \begin{vmatrix} a_{11} & \cdots & a_{1n} \\ \vdots & \ddots & \vdots \\ a_{i1} & \cdots & a_{kn} \\ \vdots & \ddots & \vdots \\ a_{k1} & \cdots & a_{kn} \\ \vdots & \ddots & \vdots \\ a_{n1} & \cdots & a_{nn} \end{vmatrix}. \tag{A.31}$$

9. If in a matrix two rows are swapped, the determinant changes its sign on opposite. We need to compare sings of the following matrices:

$$
\begin{vmatrix}
a_{11} \dots a_{1n} \\
\dots \dots \dots \\
I \\
\dots \dots \dots \\
II \\
\dots \dots \dots \\
a_{n1} \dots a_{nn}
\end{vmatrix}
\quad \text{and} \quad
\begin{vmatrix}
a_{11} \dots a_{1n} \\
\dots \dots \dots \\
II \\
\dots \dots \dots \\
I \\
\dots \dots \dots \\
a_{n1} \dots a_{nn}
\end{vmatrix} . \tag{A.32}
$$

Let us introduce a trial matrix having a zero determinant:

$$
\begin{vmatrix}
a_{11} & \dots & a_{1n} \\
\dots & \dots & \dots \\
 & I+II & \\
\dots & \dots & \dots \\
 & I+II & \\
\dots & \dots & \dots \\
a_{n1} & \dots & a_{nn}
\end{vmatrix}
=
\begin{vmatrix}
a_{11} & \dots & a_{1n} \\
\dots & \dots & \dots \\
 & I & \\
\dots & \dots & \dots \\
 & I+II & \\
\dots & \dots & \dots \\
a_{n1} & \dots & a_{nn}
\end{vmatrix}
+
\begin{vmatrix}
a_{11} & \dots & a_{1n} \\
\dots & \dots & \dots \\
 & II & \\
\dots & \dots & \dots \\
 & I+II & \\
\dots & \dots & \dots \\
a_{n1} & \dots & a_{nn}
\end{vmatrix}
= 0. \tag{A.33}
$$

From (A.33) follows

$$
\begin{vmatrix}
a_{11} \dots a_{1n} \\
\dots \dots \dots \\
I \\
\dots \dots \dots \\
I \\
\dots \dots \dots \\
a_{n1} \dots a_{nn}
\end{vmatrix}
+
\begin{vmatrix}
a_{11} \dots a_{1n} \\
\dots \dots \dots \\
I \\
\dots \dots \dots \\
II \\
\dots \dots \dots \\
a_{n1} \dots a_{nn}
\end{vmatrix}
+
\begin{vmatrix}
a_{11} \dots a_{1n} \\
\dots \dots \dots \\
II \\
\dots \dots \dots \\
I \\
\dots \dots \dots \\
a_{n1} \dots a_{nn}
\end{vmatrix}
+
\begin{vmatrix}
a_{11} \dots a_{1n} \\
\dots \dots \dots \\
II \\
\dots \dots \dots \\
II \\
\dots \dots \dots \\
a_{n1} \dots a_{nn}
\end{vmatrix}
= 0. \tag{A.34}
$$

Using property 6, we can conclude that the sum of the second and third determinants is equal to zero, since the determinants of the first and fourth matrices are zeros. This fact proves the property 9.

Appendix B
Mutual Inductance Between Two Filaments

B.1 MATLAB Functions

```
%%%%%%%%%%%%%%%%%%%%%%%%%%%%%%%%%%%%%%%%%
% Detailed explanation goes here
%%%%%%%%%%%%%%%%%%%%%%%%%%%%%%%%%%%%%%%%%
% Function returns the mutual inductance between two circular
filaments of radius Rp and Rs,
% whose centres are separated by coordinates xb,yb,zb, and angular
position
% defined by the anglar theta (tilting angle defined in
an interval of 0<=theta<=360 degrees)
%and eta (rotating angle around vertical axis defined in
an interval of 0<=theta<=360).
%The angular misalignment is defined in the same way as
given by Gover's notation).

% All dimensions must be in "meters" and angles in "radians"
%

% The units have been adapted to the S.I. system
%
% Programmed by Kirill Poletkin
%%
function M = Poletkin(Rp,Rs,xb,yb,zb,theta,eta,tol)

if nargin==7, tol=1e-10;
elseif nargin<7 || nargin >8,
    error('Wrong number of parameters in function call (Poletkin.m)!');
end
```

K. Poletkin, *Levitation Micro-Systems*, Microsystems and Nanosystems,
https://doi.org/10.1007/978-3-030-58908-0

```
if theta==pi/2 || theta==−pi/2
% Treatment of special case when Theta=90 degrees
    M0=Poletkin90(Rp,Rs,xb,yb,zb,eta,tol);

else
    a=Rs/Rp;
    dx=xb/Rs;
    dy=yb/Rs;
    dz=zb/Rs;
    M0=4e−7*sqrt(Rs*Rp)*integral(@(p)dL(p,a,dx,dy,dz,theta,eta),...
    0,2*pi,'RelTol',0,'AbsTol',tol);
end
M=M0;
```

```
%%%%%%%%%%%%%%%%%%%%%%%%%%%%%%%%%%%%%%
%%%%%%%%%%%%%%%%%%%%%%%%%%%%%%%%%%%%%%
% Returns the mutual inductance between two circular
filaments of radius Rp and Rs,
%which are mutually perpendicular to each other. Their centres are
separated
%by coordinates xb,yb,zb, and angular position
% defined by the theta =90 and the angle eta (rotating angle around
% vertical axis). (Angular misalignment is the same as given by Gover).

% All dimensions must be in "meters" and angles in "radians"
%

%
% The units have been adapted to the S.I. system
%
% Programmed by Kirill Poletkin

function M = Poletkin90(Rp,Rs,xb,yb,zb,eta,tol)


%%%%%%%%%%%%%%%%%%%%%%%%%%%%%%%%%%%
if nargin==6, tol=1e−14;
elseif nargin<6 || nargin >7,
    error('Wrong_number_of_parameters_in_function_call(Poletkin90.m)!');
end

a=Rs/Rp;
dx=xb/Rs;
dy=yb/Rs;
dz=zb/Rs;
```

```
M=4e−7*sqrt(Rs*Rp)*integral(@(r)dL(r,a,dx,dy,dz,eta,1),...
        1,−1,'RelTol',0,'AbsTol',tol)+...
    4e−7*sqrt(Rs*Rp)*integral(@(r)dL(r,a,dx,dy,dz,eta,−1),...
    −1,1,'RelTol',0,'AbsTol',tol);
%%%%%%%%%%%%%%%%%%%%%%%%%%%%%%%%%%%%%%
%%%%%%%%%%%%%%%%%%%%%%%%%%%%%%%%%%%%%%
%Integrand function
function f=dL(r,a,x,y,z,et,sg)
 % r is the length of integration
 % a is the ratio of Rs/Rp
 % th is the theta and et is the eta
 % sg is the sign for the evaluation of the hight

 %%%%%%%%%%%%%%%%%%%%%%%%%%%%%%%%%%%%%
 %%%%%% Dimensionless functions
    % dimensionless radius
    s=sqrt(x.^2+y.^2); % dimensionless parameter s
    rho=sqrt(s.^2+2.*r.*(x.*cos(et)+y.*sin(et))+r.^2);
    if sg==1, l=z+sqrt(1−r.^2); end
    if sg==−1, l=z−sqrt(1−r.^2); end
 %%%%%%%%%%%%%%%%%%%%%%%%%%%%%%%%%%%%%

 kk=4.*a.*rho./((a.*rho+1).^2+a.^2.*l.^2);
 k=sqrt(kk);

 [K,E]=ellipke(kk);

 Psi=(1−0.5*kk).*K−E;

 t1=sin(et).*(x+r.*cos(et));
 t2=cos(et).*(y+r.*sin(et));

f=(t1−t2).*Psi./(k.*rho.^1.5);
```

B.2 Determination of Angular Position of the Secondary Circular Filament

The angular position of the secondary circle can be defined through the pair of angle θ and η corresponding to manner I and the angle α and β manner II. The relationship between two pairs of angles can be determined via two spherical triangles denoted in Roman numbers I and II as shown in Fig. B.1. According to the law of sines, for spherical triangle I, we can write the following relationship:

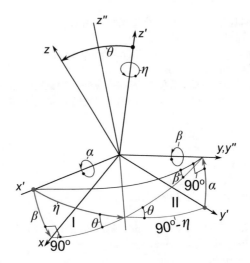

Fig. B.1 The relationship between the angles of two manners for determining angular misalignment of the secondary circle: I and II are denoted for two spherical triangles highlighted by arcs in red color

$$\frac{\sin \eta}{\sin \pi/2} = \frac{\sin \beta}{\sin \theta}.$$ (B.1)

For spherical triangle II, we have

$$\frac{\sin(\pi/2 - \eta)}{\sin(\pi/2 - \beta)} = \frac{\sin \alpha}{\sin \theta}.$$ (B.2)

Accounting for (B.1) and (B.2), the final set determining the relationship between two pairs of angles becomes

$$\begin{cases} \sin \beta = \sin \eta \sin \theta; \\ \cos \beta \sin \alpha = \cos \eta \sin \theta. \end{cases}$$ (B.3)

B.3   Presentation of Developed Formulas via the Pair of Angles α and β

Using set (B.3), we can write the following equations:

$$
\begin{cases}
\cos^2 \theta = \cos^2 \beta (1 + \sin^2 \alpha); \\
\sin^2 \theta = \sin^2 \beta + \cos^2 \beta \sin^2 \alpha; \\
\tan^2 \theta = \dfrac{\sin^2 \beta + \cos^2 \beta \sin^2 \alpha}{\cos^2 \beta (1 + \sin^2 \alpha)}; \\
\cos^2 \eta = \cos^2 \beta \sin^2 \alpha / \sin^2 \theta; \\
\sin^2 \eta = \sin^2 \beta / \sin^2 \theta.
\end{cases}
\tag{B.4}
$$

Now, applying set (B.4) to (4.17), the square of the dimensionless function $\bar{r}$ becomes

$$
\bar{r}^2 = \frac{\cos^2 \beta (1 + \sin^2 \alpha)(\sin^2 \beta + \cos^2 \beta \sin^2 \alpha)}{(\sin \varphi \cos \beta \sin \alpha - \cos \varphi \sin \beta)^2 + \cos^2 \beta (1 + \sin^2 \alpha)(\cos \varphi \cos \beta \sin \alpha + \sin \varphi \sin \beta)^2}.
\tag{B.5}
$$

Then, for the dimensionless parameter $\bar{z}_\lambda$, we have

$$
\bar{z}_\lambda = \bar{z}_B + \bar{r} \frac{\sin \varphi \cos \beta \sin \alpha - \cos \varphi \sin \beta}{\sqrt{\cos^2 \beta (1 + \sin^2 \alpha)}}.
\tag{B.6}
$$

Substituting (B.5), (B.6) and

$$
\begin{aligned}
t_1 &= \bar{x}_B + \bar{y}_B \cdot \bar{r}^2 \frac{\sin^2 \beta + \cos^2 \beta \sin^2 \alpha}{\cos^2 \beta (1 + \sin^2 \alpha)} \times \\
&\quad (\sin \varphi \cos \beta \sin \alpha - \cos \varphi \sin \beta) \times \\
&\quad (\cos \varphi \cos \beta \sin \alpha + \sin \varphi \sin \beta); \\
t_2 &= \bar{y}_B - \bar{x}_B \cdot \bar{r}^2 \frac{\sin^2 \beta + \cos^2 \beta \sin^2 \alpha}{\cos^2 \beta (1 + \sin^2 \alpha)} \times \\
&\quad (\sin \varphi \cos \beta \sin \alpha - \cos \varphi \sin \beta) \times \\
&\quad (\cos \varphi \cos \beta \sin \alpha + \sin \varphi \sin \beta),
\end{aligned}
\tag{B.7}
$$

into Eq. (4.25), the angular misalignment of the secondary circle are defined through the pair of angle α and β corresponding to the manner II.

For the case when the two circles are mutually perpendicular to each other, assuming that $\alpha = \pi/2$ then just replacing the angle η by β in formula (4.29), it can be used for calculation with new pair of angle α and β.

Appendix C
Levitation Gyroscopes

C.1 Derivation of a Levitating Gyroscope Model

In order to derive a linear model of the gyroscope, the Euler equations are used. Since an ideal levitated gyroscope can be considered as a rotating rigid body having a fixed point, and denoting the projections of the angular rate on the xyz CF as ω_x, ω_y and ω_z, respectively, then the following set can be written:

$$\left.\begin{aligned} J_x \frac{d\omega_x}{dt} + \omega_y\omega_z(J_z - J_y) &= 0; \\ J_y \frac{d\omega_y}{dt} + \omega_x\omega_z(J_x - J_z) &= 0. \end{aligned}\right\} \tag{C.1}$$

In the framework of a linear model, it is assumed that angular displacements of the rotor are small. Then, the projections of the angular rate can be defined as follows:

$$\left.\begin{aligned} \omega_x &\approx \omega_{xr} - \Omega q_y + \dot{q}_x; \\ \omega_y &\approx \omega_{yr} + \Omega q_x + \dot{q}_y; \\ \omega_z &\approx \Omega, \end{aligned}\right\} \tag{C.2}$$

where

$$\left.\begin{aligned} \omega_{xr} &= \omega_X \cos \Omega t + \omega_Y \sin \Omega t; \\ \omega_{yr} &= -\omega_X \sin \Omega t + \omega_Y \cos \Omega t, \end{aligned}\right\} \tag{C.3}$$

are the projections of the angular rate on the rotating CF. Time derivatives of ω_x and ω_y are

$$\left.\begin{aligned} \frac{d\omega_x}{dt} &\approx \dot{\omega}_{xr} - \Omega\dot{q}_y + \ddot{q}_x; \\ \frac{d\omega_y}{dt} &\approx \dot{\omega}_{yr} + \Omega\dot{q}_x + \ddot{q}_y, \end{aligned}\right\} \tag{C.4}$$

© The Editor(s) (if applicable) and The Author(s), under exclusive license to Springer Nature Switzerland AG 2021
K. Poletkin, *Levitation Micro-Systems*, Microsystems and Nanosystems, https://doi.org/10.1007/978-3-030-58908-0

where

$$
\left.\begin{array}{l}
\dfrac{d\omega_{xr}}{dt} = \dot{\omega}_X \cos \Omega t + \dot{\omega}_Y \sin \Omega t \\
\quad -\Omega\omega_X \sin \Omega t + \Omega\omega_Y \cos \Omega t; \\
\dfrac{d\omega_{yr}}{dt} = -\dot{\omega}_X \sin \Omega t + \dot{\omega}_Y \cos \Omega t \\
\quad -\Omega\omega_X \cos \Omega t - \Omega\omega_Y \sin \Omega t.
\end{array}\right\}
\tag{C.5}
$$

Substituting (C.2)–(C.5) into (C.1) and accounting for toques generated by the angular position stiffness, the linear model of the ideal levitated gyroscope becomes

$$
\left.\begin{array}{l}
J_x \ddot{q}_x + \mu_x \dot{q}_x + \left(c_s + (J_z - J_y)\Omega^2 \right) q_x \\
\quad -\Omega(J_y - J_z + J_x)\dot{q}_y \\
= \left[(J_z + J_x - J_y)\Omega \cdot \omega_X - J_y\dot{\omega}_Y \right] \sin \Omega t \\
\quad - \left[(J_z + J_x - J_y)\Omega \cdot \omega_Y + J_x\dot{\omega}_X \right] \cos \Omega t; \\
J_y \ddot{q}_y + \mu_y \dot{q}_y + \left(c_s + (J_z - J_x)\Omega^2 \right) q_y \\
\quad +\Omega(J_y - J_z + J_x)\dot{q}_x \\
= \left[(J_z - J_x + J_y)\Omega \cdot \omega_Y + J_x\dot{\omega}_X \right] \sin \Omega t \\
\quad + \left[(J_z - J_x + J_y)\Omega \cdot \omega_X - J_y\dot{\omega}_Y \right] \cos \Omega t.
\end{array}\right\}
\tag{C.6}
$$

C.2   Integral of Eq. (7.29)

In the same manner as provided in [1, p. 146], we introduce a new variable, $f = (\omega - \Omega)/\tilde{\Omega}_0$, and split the integral into four non-overlapping parts with ranges from $-\infty$ to -1, from -1 to 0, from 0 to 1, and from 1 to ∞. Hence, the integral can be represented as

$$
\begin{aligned}
&\tilde{\Omega}_0 \int_{-\infty}^{-1} + \tilde{\Omega}_0 \int_{-1}^{0} + \tilde{\Omega}_0 \int_{0}^{1} + \tilde{\Omega}_0 \int_{1}^{\infty} \\
&= \tilde{\Omega}_0 \int_{-\infty}^{-1} \frac{df/f^2 + df}{\tilde{Q}^2 \left(f - \frac{1}{f} \right)^2 + 2\tilde{Q}\frac{h}{\mu}\left(f - \frac{1}{f} \right) + 1} \\
&\quad + \Omega_0 \int_{1}^{\infty} \frac{df/f^2 + df}{\tilde{Q}^2 \left(f - \frac{1}{f} \right)^2 + 2\tilde{Q}\frac{h}{\mu}\left(f - \frac{1}{f} \right) + 1}.
\end{aligned}
\tag{C.7}
$$

Substituting $x = f - 1/f$ into (C.7), the integral becomes

$$
\Omega_0 \int_{-\infty}^{\infty} \frac{dx}{\tilde{Q}^2 x^2 + 2\tilde{Q}\frac{h}{\mu}x + 1}.
\tag{C.8}
$$

For the case of $\mu > 0$, the discriminant is

$$D = 4\tilde{Q}\frac{-\mu^2}{\tilde{\mu}^2} < 0. \tag{C.9}$$

Thus, the solution of (C.8) is [2, p. 12, Eq. 3.3.16]

$$\Omega_0 \int_{-\infty}^{\infty} = \frac{\Omega_0\tilde{\mu}}{\tilde{Q}\mu} \arctan \left.\frac{2x + 2h/\tilde{\mu}}{\mu/\tilde{\mu}}\right|_{-\infty}^{\infty} = \frac{\Omega_0\tilde{\mu} \cdot \pi}{\tilde{Q}\mu}. \tag{C.10}$$

For the case of $\mu = 0$, the discriminant is zero and integrating (C.8) also gives zero [2, p. 12, Eq. 3.3.18].

References

1. L. Brillouin, *Science and Information Theory*, 2nd edn. (Academic Press Inc., New York, 1962)
2. M. Abramowitz, I.A. Stegun, *Handbook of Mathematical Functions: with Formulas, Graphs, and Mathematical Tables*, vol. 55 (Courier Corporation, 1964)

Index

© The Editor(s) (if applicable) and The Author(s), under exclusive license
to Springer Nature Switzerland AG 2021
K. Poletkin, *Levitation Micro-Systems*, Microsystems and Nanosystems,
https://doi.org/10.1007/978-3-030-58908-0